Student Solu

for

Operations Research:

Applications and Algorithms

Fourth Edition

Wayne L. Winston
Indiana University

BROOKS/COLE
CENGAGE Learning

Australia • Brazil • Japan • Korea • Mexico • Singapore • Spain • United Kingdom • United States

BROOKS/COLE
CENGAGE Learning™

Student Solutions Manual for Operations Research: Applications and Algorithms, Fourth Edition
Wayne L. Winston

Cover Image: Getty Images, photographer
   Nick Koudis

For product information and technology assistance, contact us at
Cengage Learning Customer & Sales Support, 1-800-354-9706

For permission to use material from this text or product,
submit all requests online at **www.cengage.com/permissions**
Further permissions questions can be e-mailed to
permissionrequest@cengage.com

ISBN-13: 978-0-534-42360-5

ISBN-10: 0-534-42360-4

Brooks/Cole Cengage Learning
20 Channel Center Street
Boston, MA 02210
USA

Cengage Learning is a leading provider of customized learning solutions with office locations around the globe, including Singapore, the United Kingdom, Australia, Mexico, Brazil, and Japan. Locate your local office at **www.cengage.com/global**

Cengage Learning products are represented in Canada by
Nelson Education, Ltd.

To learn more about Brooks/Cole, visit **www.cengage.com/brookscole**

Purchase any of our products at your local college store or at our preferred online store **www.cengagebrain.com**

Printed in the United States of America
5 6 7 8 9 10 11   20 19 18 17 16

Table of Contents

Chapter 2 Solutions

Section 2.1

1a. $-A = \begin{bmatrix} -1 & -2 & -3 \\ -4 & -5 & -6 \\ -7 & -8 & -9 \end{bmatrix}$

1b. $3A = \begin{bmatrix} 3 & 6 & 9 \\ 12 & 15 & 18 \\ 21 & 24 & 27 \end{bmatrix}$

1c. A+2B is undefined.

1d. $A^T = \begin{bmatrix} 1 & 4 & 7 \\ 2 & 5 & 8 \\ 3 & 6 & 9 \end{bmatrix}$

1e. $B^T = \begin{bmatrix} 1 & 0 & 1 \\ 2 & -1 & 2 \end{bmatrix}$

1f. $AB = \begin{bmatrix} 4 & 6 \\ 10 & 15 \\ 16 & 24 \end{bmatrix}$

1g. BA is undefined.

3. Let $A = (a_{ij})$, $B = (b_{ij})$, and $C = (c_{ij})$. We must show that $A(BC) = (AB)C$. The i-j'th element of $A(BC)$ is given by

$$\sum_{x} a_{ix}\left(\sum_{y} b_{xy}c_{yj}\right) = \sum_{x}\sum_{y} a_{ix}b_{xy}c_{yj}$$

The i-j'th element of $(AB)C$ is given by

$$\sum_{y}\left(\sum_{x} a_{ix}b_{xy}\right)c_{yj} = \sum_{y}\sum_{x} a_{ix}b_{xy}c_{yj} = \sum_{x}\sum_{y} a_{ix}b_{xy}c_{yj}$$

Thus $A(BC) = (AB)C$.

Section 2.2

1. $\begin{bmatrix} 1 & -1 \\ 2 & 1 \\ 1 & 3 \end{bmatrix} \begin{bmatrix} x_1 \\ x_2 \end{bmatrix} = \begin{bmatrix} 4 \\ 6 \\ 8 \end{bmatrix}$ and $\begin{bmatrix} 1 & -1 & 4 \\ 2 & 1 & 6 \\ 1 & 3 & 8 \end{bmatrix}$

Section 2.3

1. $\begin{bmatrix} 1 & 1 & 0 & 1 & | & 3 \\ 0 & 1 & 1 & 0 & | & 4 \\ 1 & 2 & 1 & 1 & | & 8 \end{bmatrix}$ $\begin{bmatrix} 1 & 1 & 0 & 1 & | & 3 \\ 0 & 1 & 1 & 0 & | & 4 \\ 0 & 1 & 1 & 0 & | & 5 \end{bmatrix}$ $\begin{bmatrix} 1 & 0 & -1 & 1 & | & -1 \\ 0 & 1 & 1 & 0 & | & 4 \\ 0 & 0 & 0 & 0 & | & 1 \end{bmatrix}$

The last row of the last matrix indicates that the original system has no solution.

2. $\begin{bmatrix} 1 & 1 & 1 & | & 4 \\ 1 & 2 & 0 & | & 6 \end{bmatrix}$ $\begin{bmatrix} 1 & 1 & 1 & | & 4 \\ 0 & 1 & -1 & | & 2 \end{bmatrix}$ $\begin{bmatrix} 1 & 0 & 2 & | & 2 \\ 0 & 1 & -1 & | & 2 \end{bmatrix}$

This system has an infinite number of solutions of the form
$x_3 = k$, $x_1 = 2 - 2k$, $x_2 = 2 + k$.

$$\begin{array}{l} 3. \end{array} \left[\begin{array}{cc|c} 1 & 1 & 1 \\ 2 & 1 & 3 \\ 3 & 2 & 4 \end{array}\right] \left[\begin{array}{cc|c} 1 & 1 & 1 \\ 0 & -1 & 1 \\ 3 & 2 & 4 \end{array}\right] \left[\begin{array}{cc|c} 1 & 1 & 1 \\ 0 & -1 & 1 \\ 0 & -1 & 1 \end{array}\right] \left[\begin{array}{cc|c} 1 & 1 & 1 \\ 0 & 1 & -1 \\ 0 & -1 & 1 \end{array}\right] \left[\begin{array}{cc|c} 1 & 0 & 2 \\ 0 & 1 & -1 \\ 0 & 0 & 0 \end{array}\right]$$

This system has the unique solution $x_1 = 2$ $x_2 = -1$.

Section 2.4

$$\left[\begin{array}{ccc} 1 & 0 & 1 \\ 1 & 2 & 1 \\ 2 & 2 & 2 \end{array}\right] \left[\begin{array}{ccc} 1 & 0 & 1 \\ 0 & 2 & 0 \\ 2 & 2 & 2 \end{array}\right] \left[\begin{array}{ccc} 1 & 0 & 1 \\ 0 & 2 & 0 \\ 0 & 2 & 0 \end{array}\right] \left[\begin{array}{ccc} 1 & 0 & 1 \\ 0 & 1 & 0 \\ 0 & 2 & 0 \end{array}\right] \left[\begin{array}{ccc} 1 & 0 & 1 \\ 0 & 1 & 0 \\ 0 & 0 & 0 \end{array}\right]$$

Row of 0's indicates that V is linearly dependent.

$$2. \qquad \left[\begin{array}{ccc} 2 & 1 & 0 \\ 1 & 2 & 0 \\ 3 & 3 & 1 \end{array}\right] \left[\begin{array}{ccc} 1 & 1/2 & 0 \\ 1 & 2 & 0 \\ 3 & 3 & 1 \end{array}\right] \left[\begin{array}{ccc} 1 & 1/2 & 0 \\ 0 & 3/2 & 0 \\ 3 & 3 & 1 \end{array}\right] \left[\begin{array}{ccc} 1 & 1/2 & 0 \\ 0 & 3/2 & 0 \\ 0 & 3/2 & 1 \end{array}\right]$$

$$\left[\begin{array}{ccc} 1 & 1/2 & 0 \\ 0 & 1 & 0 \\ 0 & 3/2 & 1 \end{array}\right]$$

$$\left[\begin{array}{ccc} 1 & 0 & 0 \\ 0 & 1 & 0 \\ 0 & 3/2 & 1 \end{array}\right] \left[\begin{array}{ccc} 1 & 0 & 0 \\ 0 & 1 & 0 \\ 0 & 0 & 1 \end{array}\right]$$

The rank of the last matrix is 3, so V is linearly independent.

Section 2.5

$$1. \left[\begin{array}{cc|cc} 1 & 3 & 1 & 0 \\ 2 & 5 & 0 & 1 \end{array}\right] \left[\begin{array}{cc|cc} 1 & 3 & 1 & 0 \\ 0 & -1 & -2 & 1 \end{array}\right] \left[\begin{array}{cc|cc} 1 & 3 & 1 & 0 \\ 0 & 1 & 2 & -1 \end{array}\right] \left[\begin{array}{cc|cc} 1 & 0 & -5 & 3 \\ 0 & 1 & 2 & -1 \end{array}\right]$$

3

Thus $A^{-1} = \begin{bmatrix} -5 & 3 \\ 2 & -1 \end{bmatrix}$

3. $\left[\begin{array}{ccc|ccc} 1 & 0 & 1 & 1 & 0 & 0 \\ 1 & 1 & 1 & 0 & 1 & 0 \\ 2 & 1 & 2 & 0 & 0 & 1 \end{array}\right] \left[\begin{array}{ccc|ccc} 1 & 0 & 1 & 1 & 0 & 0 \\ 0 & 1 & 0 & -1 & 1 & 0 \\ 2 & 1 & 2 & 0 & 0 & 1 \end{array}\right] \left[\begin{array}{ccc|ccc} 1 & 0 & 1 & 1 & 0 & 0 \\ 0 & 1 & 0 & -1 & 1 & 0 \\ 0 & 1 & 0 & -2 & 0 & 1 \end{array}\right]$

$\left[\begin{array}{ccc|ccc} 1 & 0 & 1 & 1 & 0 & 0 \\ 0 & 1 & 0 & -1 & 1 & 1 \\ 0 & 0 & 0 & -1 & -1 & 1 \end{array}\right]$

Since we can never transform what is to the left of $|$ into I_3, A^{-1} does not exist.

Section 2.6

1. Expansion by row 2 yields

$(-1)^{2+1}(4)(-6) + (-1)^{2+2}(5)(-12) + (-1)^{2+3}6(-6) = 0$
  Expansion by row 3 yields

$(-1)^{3+1}(7)(-3) + (-1)^{3+2}(8)(-6) + (-1)^{3+3}(9)(-3) = 0$

Chapter 3 Solutions

Section 3.1

1. max $z = 30x_1 + 100x_2$
       s.t. $x_1 + x_2 \leq 7$ (Land Constraint)
              $4x_1 + 10x_2 \leq 40$(Labor Constraint)
              $10x_1 \geq 30$(Govt. Constraint)
              $x_1 \geq 0, x_2 \geq 0$

3. 1 bushel of corn uses 1/10 acre of land and 4/10 hours of labor while 1 bushel of wheat uses 1/25 acre of land and 10/25 hours of labor. This yields the following formulation:

max $z = 3x_1 + 4x_2$
s.t. $x_1/10 + x_2/25 \leq 7$ (Land Constraint)
         $4x_1/10 + 10x_2/25 \leq 40$ (Labor Const.)
         $x_1 \geq 30$ (Govt. Const.)
         $x_1 \geq 0, x_2 \geq 0$

Section 3.2

1. EF is $4x_1 + 10x_2 = 40$, CD is $x_1 = 3$, and AB is $x_1 + x_2 = 7$. The feasible region is bounded by ACGH. The dotted line in graph is isoprofit line $30x_1 + 100x_2 = 120.$ Point G is optimal. At G the constraints $10x_1 \geq 30$ and $4x_1 + 10x_2 \leq 40$ are binding. Thus optimal solution has $x_1 = 3$, $x_2 = 2.8$ and $z = 30(3) + 100(2.8) = 370$.

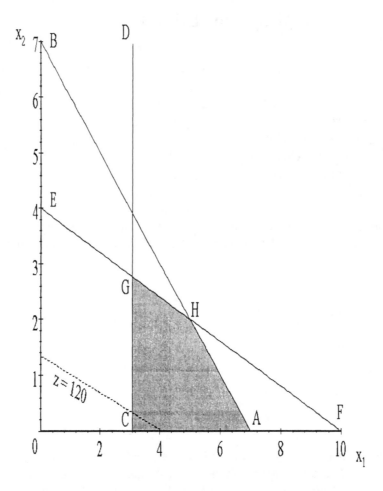

3. x_1 = Number of hours of Process 1 and x_2 = Number of hours of Process 2. Then the appropriate LP is

min $z = 4x_1 + x_2$

s.t. $3x_1 + x_2 \geq 10$ (A constraint)

  $x_1 + x_2 \geq 5$ (B constraint)

  $x_1 \geq 3$ (C constraint)

  $x_1\ x_2 \geq 0$

AB is $3x_1 + x_2 = 10$. CD is $x_1 + x_2 = 5$. EF is $x_1 = 3$. The feasible region is shaded. Dotted line is isocost line $4x_1 + x_2 = 24$. Moving isocost line down to left we see that H (where B and C constraints intersect) is optimal. Thus optimal solution to LP is $x_1 = 3$, $x_2 = 2$, $z = 4(3) + 2 = \$14$.

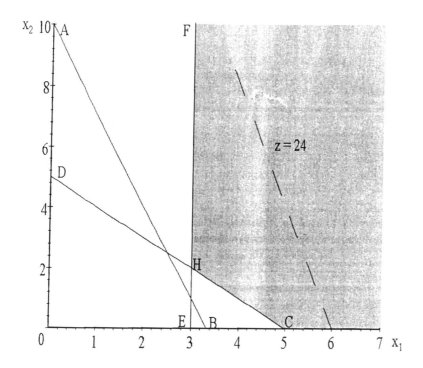

Section 3.3

1. AB is $x_1 + x_2 = 4$. CD is $x_1 - x_2 = 5$. From graph we see that there is no feasible solution (Case 3).

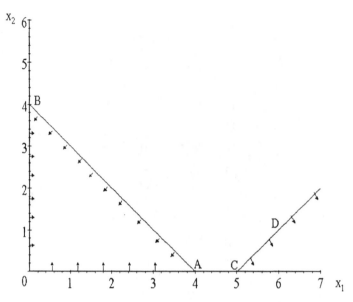

2. AB is $8x_1 + 2x_2 = 16$. CD is $5x_1 + 2x_2 = 12$. Dotted line

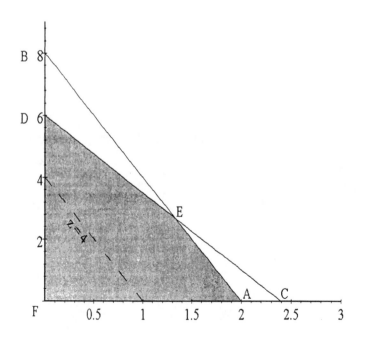

is $z = 4x_1 + x_2 = 4$. Feasible region is bounded by AEDF. Since isoprofit line is parallel to AE, entire line segment AE is optimal. Thus we have alternative or multiple optimal solutions.

3. AB is $x_1 - x_2 = 4$. AC is $x_1 + 2x_2 = 4$. Feasible region is bounded by AC and infinite line segment AB. Dotted line is isoprofit line $z = 0$. To increase z we move parallel to isoprofit line in an upward and 'leftward' direction. We will never entirely lose contact with the feasible region, so we have an unbounded LP (Case 4).

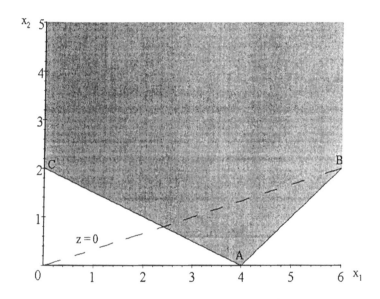

Section 3.4

1. For i = 1, 2, 3 let x_i = Tons of processed Factory i waste.
Then appropriate LP is

min $z = 15x_1 + 10x_2 + 20x_3$
s.t. $.10x_1 + .20x_2 + .40x_3 \geq 30$(Pollutant 1)
      $.45x_1 + .25x_2 + .30x_3 \geq 40$(Pollutant 2)
      $x_1 \geq 0, x_2 \geq 0, x_3 \geq 0$

It is doubtful that the processing cost is proportional to the amount of waste processed. For example, processing 10 tons of waste is probably not ten times as costly as processing 1 ton of waste. The divisibility and certainty assumptions seem reasonable.

Section 3.5

2. Let x_1 = employees starting at midnight
   x_2 = employees starting at 4 AM
   x_3 = employees starting at 8 AM
   x_4 = employees starting at noon
   x_5 = employees starting at 4 PM
   x_6 = employees starting at 8 PM
   Then a correct formulation is

$$\min z = \quad x_1 + x_2 + x_3 + x_4 + x_5 + x_6$$

s.t. $\quad x_1 + x_6 \geq 8$

$$x_1 + x_2 \geq 7$$

$$x_2 + x_3 \geq 6$$
$$x_3 + x_4 \geq 6$$
$$x_4 + x_5 \geq 5$$
$$x_5 + x_6 \geq 4$$
$$x_1, x_2, x_3, x_4, x_5, x_6 \geq 0$$

Section 3.6

2. NPV of Investment $1 = -6 - 5/1.1 + 7/(1.1)^2 + 9/(1.1)^3 = \$2.00$
   NPV of Investment $2 = -8 - 3/(1.1) + 9/(1.1)^2 + 7/(1.1)^3 = \$1.97$
Let x_1 = Fraction of Investment 1 that is undertaken and
   x_2 = Fraction of Investment 2 that is undertaken. If we measure NPV
in thousands of dollars, we wish to solve the following LP:

$$\max z = 2x_1 + 1.97x_2$$

s.t. $\quad 6x_1 + 8x_2 \leq 10$

$$5x_1 + 3x_2 \leq 7$$

$$x_1 \leq 1$$

$$x_2 \leq 1$$

From the following graph we find the optimal solution to this LP to be $x_1 = 1$, $x_2 = .5$, $z = \$2,985$.

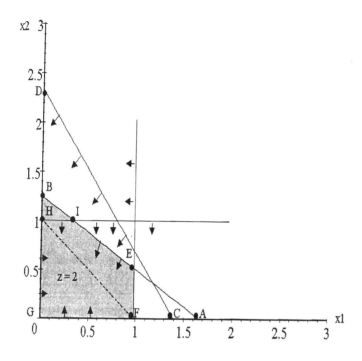

Section 3.8

1. Let (all variables are in ounces) Ing. 1 = Sugar, Ing. 2 = Nuts, Ing. 3 = Chocolate, Candy 1 = Slugger and Candy 2 = Easy Out Let x_{ij} = Ounces of Ing. i used to make candy j.

The appropriate LP is

max z = $25(x_{12} + x_{22} + x_{32}) + 20(x_{11} + x_{21} + x_{31})$

s.t. $x_{11} + x_{12} \leq 100$ (Sugar Const.)

$x_{21} + x_{22} \leq 20$ (Nuts Constraint)

$x_{31} + x_{32} \leq 30$ (Chocolate Const.)

(1) $x_{22}/(x_{12} + x_{22} + x_{32}) \geq .20$

(2) $x_{21}/(x_{11} + x_{21} + x_{31}) \geq .10$

(3) $x_{31}/(x_{11} + x_{21} + x_{31}) \geq .10$

(1) -(3) are not LP constraints and should be replaced by the following three constraints:

Replace (1) by $x_{22} \geq .2(x_{12}+x_{22}+x_{32})$ or $.8x_{22}-.2x_{12}-.2x_{32} \geq 0$

Replace (2) by $x_{21} \geq .1(x_{11}+x_{21}+x_{31})$ or $.9x_{21}-.1x_{11}-.1x_{31} \geq 0$

Replace (3) by $x_{31} \geq .1(x_{11}+x_{21}+x_{31})$ or $.9x_{31}-.1x_{11}-.1x_{21} \geq 0$

5. xij = barrels of oil i used to make product j (j = 1 is gasoline and j = 2 is heating oil)

ai = dollars spent advertising product i

max z = $25(x11 + x21) + 20(x12 + x22) - a1 - a2$

11

s.t. x11 + x21 = 5a1, x12 + x22 = 10a2, 2x11 - 3x21≥0 x11 + x12≤5000 x21 + x22≤10,000

4x12 - x22≥0 All variables ≥0.

Section 3.9

1. Let x_1 = Hours of Process 1 run per week.

x_2 = Hours of Process 2 run per week.

x_3 = Hours of Process 3 run per week.

g_2 = Barrels of Gas 2 sold per week.

o_1 = Barrels of Oil 1 purchased per week

o_2 = Barrels of Oil 2 purchased per week

max z = $9(2x_1)+10g_2+24(2x_3)-5x_1-4x_2-x_3-2o_1-3o_2$

s.t. $o_1 = 2x_1 +x_2$

$o_2 = 3x_1+3x_2+2x_3$

$o_1 ≤ 200$

$o_2 ≤ 300$

$g_2+3x_3 = x_1+3x_2$ (Gas 2 Prod.)

$x_1+x_2+x_3 ≤ 100$ (100 hours per wk. of cracker   time)

All variables ≥0.

3. Let x6 = pounds of raw material used to produce   Brute

x7 = pounds of raw material used to produce Chanelle

Then the appropriate formulation is:

max z = 7x1 + 14x2 + 6x3 + 10x4 - 3x5

s.t.   x5 ≤ 4000

3x2 + 2x4 + x5 ≤ 6000

x1 + x2 = 3x6

x3 + x4 = 4x7

x5 = x6 + x7

All variables ≥0

Section 3.10

1. Let x_t = Production during month t and i_t = Inventory at end of month t.

min z   = $5x_1+8x_2+4x_3+7x_4+2i_1+2i_2+2i_3+2i_4-6i_4$

s.t. $i_1 = x_1 - 50$

$i_2 = i_1 + x_2 - 65$

$i_3 = i_2 + x_3 - 100$

$i_4 = i_3 + x_4 - 70$

All variables ≥0.

3. Let Ct = cheesecakes baked during month t
   Bt = black forest cakes baked during month t
   It = number of cheesecakes in inventory at end of month t
   It'= number of black forest cakes in inventory at the end of month t.
   Then an appropriate formulation is

$$\min z = 3C1 + 3.4C2 + 3.8C3 + 2.5B1 + 2.8B2 + 3.4B3$$
$$+.5(I1 + I2 +I3) + .4(I1' + I2' + I3')$$

s.t.
$$C1 + B1 \leq 65$$
$$C2 + B2 \leq 65$$
$$C3 + B3 \leq 65$$
$$I1 = C1 - 40$$
$$I2 = I1 + C2 - 30$$
$$I3 = I2 + C3 - 20$$
$$I1' = B1 - 20$$
$$I2' = I1' + B2 - 30$$
$$I3' = I2' + B3 - 10$$
All variables ≥ 0

Section 3.11

3. Let A = $ invested in A.
   B = $ invested in B.
   c_0 = Leftover cash at Time 0
   c_1 = Leftover cash at Time 1
   c_2 = Leftover cash at Time 2.
Then a correct formulation is
$\max z = c_2 + 1.9B$
s.t. $A + c_0 = 10,000$ (Time 0 Avail = Time 0 Inv.)
   $.2A + c_0 = B + c_1$ (Time 1 Avail. = Time 1 Inv.)
   $1.5A + c_1 = c_2$ (Time 2 Avail. = Time 2 Inv.)
All Variables ≥ 0.

The optimal solution to this LP is $B = c_0 = \$10,000$,
$A = c_1 = c_2 = 0$ and $z = \$19,000$. Notice that it is optimal to `wait' for the `good'
investment (B) even though leftover cash earns no interest.

Section 3.12

2. Let JAN1 = Number of computers rented at beginning of JAN for one month, etc. Also define IJAN = Number of computers available to meet January demand, etc. The appropriate LP is

$$\min z = 200(JAN1 + FEB1 + MAR1 + APR1 + MAY1 + JUNE1) + 350(JAN2 + FEB2 + MAR2 + APR2$$
$$+MAY2+JUN2)+450(JAN3+FEB3+MAR3+APR3+MAY3+JUN3)-150MAY3-300JUN3$$
$$-175JUN2$$

s.t.   IJAN = JAN1 + JAN2 + JAN3
   IFEB = IJAN - JAN1 + FEB1 + FEB2 + FEB3
   IMAR = IFEB - JAN2 - FEB1 + MAR1 + MAR2 + MAR3
   IAPR = IMAR - FEB2 - MAR1 - JAN3 + APR1 + APR2 + APR3
   IMAY = IAPR - FEB3 - MAR2 - APR1 + MAY1 + MAY2 + MAY3
   IJUN = IMAY - MAR3 - APR2 - MAY1 + JUN1 + JUN2 + JUN3
   IJAN$\geq$9
   IFEB$\geq$5
   IMAR$\geq$7
   IAPR$\geq$9
   IMAY$\geq$10
   IJUN$\geq$5   All variables $\geq$0

Chapter 4 Solutions

Section 4.1

1. $\max z = 3x_1 + 2x_2$
    s.t. $2x_1 + x_2 + s_1 = 100$
    $ x_1 + x_2 + s_2 = 80$
    $ x_1 + s_3 = 40$

Section 4.4

1. From Figure 2 of Chapter 3 we see that the extreme points of the feasible region are

    Basic Feasible Solution

    $H = (0, 0)$ $s_1 = 100, s_2 = 80, s_3 = 40$ $x_1 = x_2 = x_3 = 0$
    $E = (40, 0)$ $x_1 = 40, s_1 = 20, s_2 = 40$ $x_2 = x_3 = s_3 = 0$
    $F = (40, 20)$ $x_1 = 40, x_2 = 20, s_2 = 20$ $x_3 = s_1 = s_3 = 0$
    $G = (20, 60)$ $x_1 = 20, x_2 = 60, s_3 = 20$ $x_3 = s_1 = s_2 = 0$
    $D = (0, 80)$ $s_1 = 20, x_2 = 80, s_3 = 40$ $s_2 = x_1 = x_3 = 0$

Section 4.5

2.

z	x_1	x_2	s_1	s_2	RHS	Ratio
1	-2	-3	0	0	0	
0	1	2	1	0	6	3* Enter x_2 in row 1
0	2	1	0	1	8	8
0	-1/2	0	3/2	0	9	
0	1/2	1	1/2	0	3	6
0	3/2	0	-1/2	1	5	10/3* Enter x_1 in row 2

z	X₁	X₂	S₁	S₂	RHS	Ratio

$$z \quad x_1 \quad x_2 \quad s_1 \quad s_2 \quad \text{RHS} \quad \text{Ratio}$$

0	0	0	4/3	1/3	32/3

0	0	1	2/3	-1/3	4/3

0	1	0	-1/3	2/3	10/3

This is an optimal tableau and the optimal solution is $z = 32/3$, $x_1 = 10/3$, $x_2 = 4/3$, $s_1 = s_2 = 0$.

Section 4.6

3.

Z	X1	X2	S1	S2	RHS
1	-2	5	0	0	0
0	3	8	1	0	12
0	2	3	0	1	6

We enter X2 into the basis in Row 1. The resulting optimal tableau is

Z	X1	X2	S1	S2	RHS
1	-31/8	0	-5/8	0	-7.5
0	3/8	1	1/8	0	1.5
0	7/8	-7/3	-3/8	1	1.5

The LP's optimal solution is Z= -7.5 , X1=0, and X2=1.5.

Section 4.7

5.

Z	X1	X2	S1	S2	RHS
1	-2	-2	0	0	0
0	1	1	1	0	6
0	2	1	0	1	13

We now arbitrarily choose to enter X1 into basis. X1 enters basis in Row 1 yielding following optimal tableau.

Z	X1	X2	S1	S2	RHS
1	0	0	2	0	12
0	1	1	1	0	6
0	0	-1	-2	1	1

This tableau yields the optimal solution Z = 12, X1 = 6, X2=0. Pivoting X2 into the basis yields the alternative optimal solution Z=12, X1=0, X2=6. All optimal solutions are of the form c(1st optimal solution) + $(1-c)$(2nd optimal solution) where $0<=c<=1$. This shows all optimal solutions are of form Z=12, X1= 6c, X2=6-6c, $0<=c<=1$.

Section 4.8

6.

Z	X1	X2	S1	S2	RHS
1	1	3	0	0	0
0	1	-2	1	0	4
0	-1	1	0	1	3

X2 enters the basis in ROW 2 yielding the following tableau:

Z	X1	X2	S1	S2	RHS
1	4	0	0	-3	-9
0	-1	0	1	2	10
0	-1	1	0	1	3

We would like to enter X1 into the basis but there is no row in which X1 has a positive coefficient. Therefore the LP is unbounded.

Section 4.10

4. MODEL:
SETS:
PRODUCTS/1..3/:MADE,PROFIT;
RESOURCES/1..3/:AVAIL;
RESPRO(RESOURCES,PRODUCTS):USAGE;
ENDSETS
MAX=@SUM(PRODUCTS(I):PROFIT(I)*MADE(I));
@FOR(RESOURCES(I):@SUM(PRODUCTS(J):MADE(J)*USAGE(I,J))<=AV
AIL(I));
DATA:
PROFIT= 800,1500,2500;
AVAIL=50,10 150;
USAGE= 2,3,5

.3,.7,.2
       120,12,20;
ENDDATA
END
Optimal Solution is
Global optimal solution found at step: 1
   Objective value: 18750.00

Variable	Value	Reduced Cost
MADE(1)	0.0000000	14200.00
MADE(2)	0.0000000	0.0000000
MADE(3)	7.500000	0.0000000

Section 4.11

4. Here are the pivots:

Z	X1	X2	X3	X4	S1	S2	S3	RHS
1	3	-1	6	0	0	0	0	0
0	9	1	-9	-2	1	0	0	0
0	1	1/3	-2	-1/3	0	1	0	0
0	-9	-1	9	2	0	0	1	1

X2 now enters in Row 1 yielding the following tableau.

Z	X1	X2	X3	X4	S1	S2	S3	RHS
1	12	0	-3	-2	1	0	0	0
0	9	1	-9	-2	1	0	0	0
0	-2	0	1	1/3	-1/3	1	0	0
0	0	0	0	0	1	0	1	1

We now enter X3 into the basis in Row 2.

Z	X1	X2	X3	X4	S1	S2	S3	RHS
1	6	0	0	-1	0	3	0	0
0	-9	1	0	1	-2	9	0	0
0	-2	0	1	1/3	-1/3	1	0	0
0	0	0	0	0	1	0	1	1

We now enter X4 into the basis and arbitrarily choose to enter X4 in Row
1.

18

Z	X1	X2	X3	X4	S1	S2	S3	RHS
1	-3	1	0	0	-2	12	0	0
0	-9	1	0	1	-2	9	0	0
0	1	-1/3	1	0	1/3	-2	0	0
0	0	0	0	0	1	0	1	1

X1 now enters basis in Row 2.

Z	X1	X2	X3	X4	S1	S2	S3	RHS
1	0	0	3	0	-1	6	0	0
0	0	-2	9	1	1	-9	0	0
0	1	-1/3	1	0	1/3	-2	0	0
0	0	0	0	0	1	0	1	1

We now choose to enter S1 in Row 1.

Z	X1	X2	X3	X4	S1	S2	S3	RHS
1	0	-2	12	1	0	-3	0	0
0	0	-2	9	1	1	-9	0	0
0	1	1/3	-2	-1/3	0	1	0	0
0	0	2	-9	-1	0	9	1	1

S2 would now enter basis in Row 2. This will bring us back to initial tableau, so cycling has occurred.

Section 4.12

5. Initial tableau is

Z	X1	X2	X3	A1	A2	RHS
1	3M-1	2M-1	3M	0	0	6M
0	2	1	1	1	0	4
0	1	1	2	0	1	2

We now enter X3 in the basis in row 2:

Z	X1	X2	X3	A1	A2	RHS
1	3M/2-1	M/2-1	0	0	-3M/2	3M
0	3/2	1/2	0	1	-1/2	3
0	1/2	1/2	1	0	1/2	1

We now enter X1 in the basis in row 2.

Z	X1	X2	X3	A1	A2	RHS
1	0	-M	2-3M	0	1-3M	2
0	0	-1	-3	1	-2	0
0	1	1	2	0	1	2

This optimal tableau yields the optimal solution Z=2, X1=2, X2=X3=0.

Section 4.13

5. Initial Phase I Tableau is

W	X1	X2	X3	A1	A2	RHS
1	3	2	3	0	0	6
0	2	1	1	1	0	4
0	1	1	2	0	1	2

X1 enters the basis in Row 1 yielding

W	X1	X2	X3	A1	A2	RHS
1	0	1/2	3/2	-3/2	0	0
0	1	1/2	1/2	1/2	0	2
0	0	1/2	3/2	-1/2	1	0

This completes Phase I. This is Case III so we now drop the A1 column and begin Phase II with A2 as a basic variable.

Z	X1	X2	X3	A2	RHS
1	0	-1/2	1/2	0	2
0	1	1/2	1/2	0	2
0	0	1/2	3/2	1	0

X2 now enters basis in Row 2 yielding the following optimal tableau:

Z	X1	X2	X3	A2	RHS
1	0	0	2	1	2
0	1	0	-1	-1	2
0	0	1	3	2	0

The optimal solution is Z=2, X1=2, X2=X3=0.

Section 4.14

1. Let $i_t = i_t' - i_t''$ be the inventory position at the end of month t. For each constraint in original problem replace i_t by $i_t' - i_t''$. Also add the sign restrictions $i_t' >= 0$ and $i_t'' >= 0$. To ensure that demand is met by end of

Quarter 4 add sign restriction $i_4' - i_4'' \geq 0$. Change the terms involving i_t in objective function to $100i_1' + 110i_1'' + 100i_2' + 110i_2'' + 100i_3' + 110i_3'' + 100i_4' + 110i_4''$.

Section 4.16

1a.   The only point satisfying the LIP constraint and the budget constraint is (6, 0). Thus $x_1 = 6$, $x_2 = 0$ is the optimal solution. This leaves the HIW goal unmet by 5 million.

1b.   Only point satisfying the HIM constraint, the budget line, and the LIP constraint, is the point (6,0). Thus

        $x_1 = 6$, $x_2 = 0$ is the optimal solution.

1c.   The point satisfying both the budget constraint and the HIM(highest priority goal)goal that is closest to the HIW constraint occurs where the HIM and budget lines intersect. This is at the point $x_1 = 5$ and $x_2 = 5/3$.

1d.   The point satisfying the budget constraint and HIW that is closest to HIM is C. Thus $x_1 = 3$ and $x_2 = 5$ is the optimal solution.

5.   Let H = pounds of head in mixture
        C = pounds of chuck in mixture
        MU = pounds of mutton in mixture
        MO = pounds of moisture in mixture
Then the appropriate goal programming   formulation is

$$\min P_1 s_1^- + P_2 s_2^+ + P_3 s_3^+$$

st. $2H + .26C + .08MU + s_1^- - s_1^+ = 15$ (protein)
       $.05H + .24C + .11MU + s_2^- - s_2^+ = 8$ (fat)
       $12H + 9C + 8MU + s_3^- - s_3^+ = 800$ (cost)
       $H + C + MU + MO = 100$
       All variables nonnegative

Section 4.17

1. See file S4_17_1.xls

	A	B	C	D	E	F	G
1	SECTION 4-17						
2	PROBLEM1						
3		SUPP1	SUPP2	SUPP3	TOTALCOST		
4	AMOUNT	1200	0	100	6300		
5	COST	5	4	3	BOUGHT		NEEDED
6	LARGE	0.4	0.3	0.2	500	=>=	500
7	MEDIUM	0.4	0.35	0.2	500	>=	300
8	SMALL	0.2	0.35	0.6	300	=>=	300

2.

	A	B	C	D	E	F	G	H	I
1	PROBLEM2								
2	SECTION 4-17								
3									
4									
5	START		MON	TUES	WED	THUR	FRI	SAT	SUN
6	MON	6.3333333	1	1	1	1	1	0	0
7	TUES	5	0	1	1	1	1	1	0
8	WED	0.3333333	0	0	1	1	1	1	1
9	THURS	7.3333333	1	0	0	1	1	1	1
10	FRI	0	1	1	0	0	1	1	1
11	SAT	3.3333333	1	1	1	0	0	1	1
12	SUN	0	1	1	1	1	0	0	1
13	TOTAL	22.333333							
14									
15		AVAILABLE	17	14.667	15	19	19	16	11
16			=>=	>=	=>=	=>=	>=	=>=	=>=
17		NEEDED	17	13	15	19	14	16	11

See file S4_17_2.xls

23

Chapter 5 Solutions

Section 5.1

1. Typical isoprofit line is $3x_1+c_2x_2=z$. This has slope $-3/c_2$. If slope of isoprofit line is <-2, then Point C is optimal. Thus if $-3/c_2<-2$ or $c_2<1.5$ the current basis is no longer optimal. Also if the slope of the isoprofit line is >-1 Point A will be optimal. Thus if $-3/c_2>-1$ or $c_2>3$ the current basis is no longer optimal. Thus for $1.5\leq c_2\leq3$ the current basis remains optimal.
For $c_2 = 2.5$ $x_1 = 20$, $x_2 = 60$, but $z = 3(20) + 2.5(60) = \$210$.

2. Currently Number of Available Carpentry Hours $= b_2 = 80$. If we reduce the number of available carpentry hours we see that when the carpentry constraint moves past the point $(40, 20)$ the carpentry and finishing hours constraints will be binding at a point where $x_1>40$. In this situation $b_2<40 + 20 = 60$. Thus for $b_2<60$ the current basis is no longer optimal. If we increase the number of available carpentry hours we see that when the carpentry constraint moves past $(0, 100)$ the carpentry and finishing hours constraints will both be binding at a point where $x_1<0$. In this situation $b_2>100$. Thus if $b_2>100$ the current basis is no longer optimal. Thus the current basis remains optimal for $60\leq b_2\leq100$. If $60\leq b_2\leq100$, the number of soldiers and trains produced will change.

Section 5.2

1a. $4250 - 5(75) = \$3875$. Note we are in allowable range because land can decrease by up to 6.667 acres.

1b. Optimal solution is still to plant 25 acres of wheat and 20 acres of corn and use 350 labor hours.
New Profit $= 130(25) + 200(20) - 10(350) = \$3750$ or New Profit=Old Profit -20 (25) $= \$3750$ (This follows because profit coefficient for AI is decreased by 20 and AD=30)

1c. Allowable Decrease $=$ SLACK $=15$ and Allowable Increase $=$ Infinity. 130 bushels is in allowable range so solution remains unchanged.

2a. Decision variables remain the same.
New z-Value $=$ Old z-value$+10(88) = \$33,420$

2b. Relevant Shadow Price is -$20. Current basis remains optimal if demand is decreased by up to 3 cars, so Dual Price may be used to compute new z-value.
   New Profit = $32,540 + (-2) (-20) = $33,580.

Section 5.3

2. Cannot answer this question because current basis is no longer optimal if one more machine is available.

3. If you were given an extra ounce of chocolate, this would reduce cost by 2.5 cents, so you would be willing to pay up to 2.5 cents for an ounce of chocolate.

4. Plant 1 shadow price is $2, so another unit of capacity at Plant 1 reduces cost by $2. Thus company can pay up to $2 for an extra unit of capacity at Plant 1 and still be better off.

Section 5.4

3.   Optimal Solution   Optimal z-value
      $0 \leq c_2 \leq 1.5$ $x_1 = 40$ $x_2 = 20$ $120 + 20c_2$
      $1.5 \leq c_2 \leq 3$ $x_1 = 20$ $x_2 = 60$ $60 + 60c_2$
      $c_2 \geq 3$ $x_1 = 0$ $x_2 = 80$ $80c_2$

Also see graphs on the following pages.

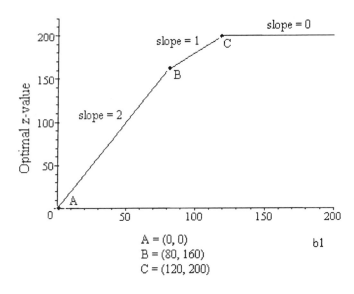

A = (0, 0)
B = (80, 160)
C = (120, 200)

b1

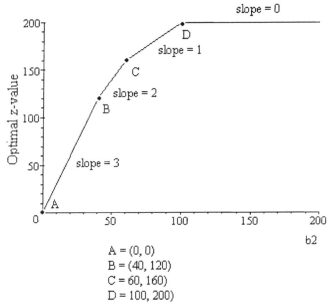

A = (0, 0)
B = (40, 120)
C = 60, 160)
D = 100, 200)

b2

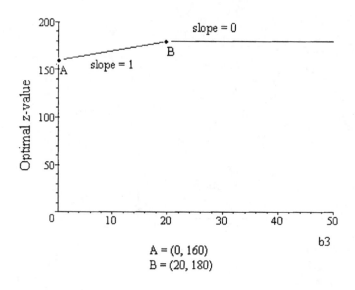

A = (0, 160)
B = (20, 180)

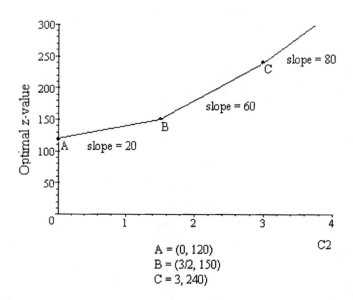

A = (0, 120)
B = (3/2, 150)
C = 3, 240)

Chapter 6 Solutions

Section 6.1

1. Typical isoprofit line is $3x_1 + c_2 x_2 = z$. This has slope $-3/c_2$. If slope of isoprofit line is <-2, then Point C is optimal. Thus if $-3/c_2 < -2$ or $c_2 < 1.5$ the current basis is no longer optimal. Also if the slope of the isoprofit line is >-1 Point A will be optimal. Thus if $-3/c_2 > -1$ or $c_2 > 3$ the current basis is no longer optimal. Thus for $1.5 \leq c_2 \leq 3$ the current basis remains optimal.
For $c_2 = 2.5$ $x_1 = 20$, $x_2 = 60$, but $z = 3(20) + 2.5(60) = \$210$.

2. Currently Number of Available Carpentry Hours $= b_2 = 80$. If we reduce the number of available carpentry hours we see that when the carpentry constraint moves past the point $(40, 20)$ the carpentry and finishing hours constraints will be binding at a point where $x_1 > 40$. In this situation $b_2 < 40 + 20 = 60$. Thus for $b_2 < 60$ the current basis is no longer optimal. If we increase the number of available carpentry hours we see that when the carpentry constraint moves past $(0, 100)$ the carpentry and finishing hours constraints will both be binding at a point where $x_1 < 0$. In this situation $b_2 > 100$. Thus if $b_2 > 100$ the current basis is no longer optimal. Thus the current basis remains optimal for $60 \leq b_2 \leq 100$. If $60 \leq b_2 \leq 100$, the number of soldiers and trains produced will change.

Section 6.2

1. $BV = \{x_1, x_2\}$ $B = \begin{bmatrix} 2 & -1 \\ -1 & 1 \end{bmatrix}$ $B^{-1} = \begin{bmatrix} 1 & 1 \\ 1 & 2 \end{bmatrix}$ $c_{BV} = [3 \ 1]$

$c_{BV}B^{-1} = [4 \ 5]$
Coefficient of s_1 in row $0 = 4$
Coefficient of s_2 in row $0 = 5$

Right hand Side of row $0 = c_{BV}B^{-1}b = [4 \ 5] \begin{bmatrix} 2 \\ 4 \end{bmatrix} = 28$

s_1 column $= B^{-1} \begin{bmatrix} 1 \\ 0 \end{bmatrix} = \begin{bmatrix} 1 \\ 1 \end{bmatrix}$

$$s_2 \text{ column} = B^{-1} \begin{bmatrix} 0 \\ 1 \end{bmatrix} = \begin{bmatrix} 1 \\ 2 \end{bmatrix}$$

$$x_1 \text{ column} = \begin{bmatrix} 1 \\ 0 \end{bmatrix} \quad x_2 \text{ column} = \begin{bmatrix} 0 \\ 1 \end{bmatrix}$$

$$\text{Right hand Side of Constraints} = B^{-1}b = \begin{bmatrix} 1 & 1 \\ 1 & 2 \end{bmatrix} \begin{bmatrix} 2 \\ 4 \end{bmatrix} = \begin{bmatrix} 6 \\ 10 \end{bmatrix}$$

Thus the optimal tableau is
$$z + 4s_1 + 5s_2 = 28$$
$$x_1 + s_1 + s_2 = 6$$
$$x_2 + s_1 + 2s_2 = 10$$

Section 6.3

6a. x_1 is non-basic so changing the coefficient of x_1 in the objective function will only change the coefficient of x_1 in the optimal row 0. Let the new coefficient of x_1 in the objective function be $3 + \Delta$. The new coefficient of x_1 in the optimal row 0 will be
$c_{BV}B^{-1}a_1 - (3 + \Delta) = 3 - \Delta$. Thus if $3 - \Delta \geq 0$ or $\Delta \leq 3$ the current basis remains optimal. Thus if profit for a Type 1 Candy Bar is ≤ 6 cents the current basis remains optimal.

6b. Changing Candy Bar 2 profit to $7 + \Delta$ changes $c_{BV}B^{-1}$ to
$$[5 \; 7 + \Delta] \begin{bmatrix} 3/2 & -1/2 \\ -1/2 & 1/2 \end{bmatrix} = [4 - \Delta/2 \; 1 + \Delta/2]$$

$$\text{Then coefficient of } x_1 \text{ in row } 0 = [4 - \Delta/2 \; 1 + \Delta/2] \begin{bmatrix} 1 \\ -3 \\ 2 \end{bmatrix} = 3 + \Delta/2.$$

Thus row 0 of optimal tableau is now
$$z + (3 + \Delta/2) x_1 + (4 - \Delta/2) s_1 + (1 + \Delta/2)s_2 = ?$$

Thus current basis remains optimal if (1)-(3) are met:
(1) $3 + \Delta/2 \geq 0$ (or $\Delta \geq -6$)

30

(2) $4 - \Delta/2 \geq 0$ (or $\Delta \leq 8$)

(3) $1 + \Delta/2 \geq 0$ (or $\Delta \geq -2$)

Thus if $-2 \leq \Delta \leq 8$ the current basis remains optimal. Thus if profit for Type 2 Candy Bar is between $7 - 2 = 5$ and $7 + 8 = 15$ cents the current basis remains optimal.

6c. If the amount of sugar available is changed to $50 + \Delta$ the current basis remains optimal iff

$$\begin{bmatrix} 3/2 & -1/2 \\ -1/2 & 1/2 \end{bmatrix} \begin{bmatrix} 50 & + & \Delta \\ & 100 & \end{bmatrix} = \begin{bmatrix} 25 & + & 3\Delta/2 \\ 25 & - & \Delta/2 \end{bmatrix}$$

Thus current basis remains optimal iff (1) - (2) hold

(1) $25 + 3\Delta/2 \geq 0$ (or $\Delta \geq -50/3$)

(2) $25 - \Delta/2 \geq 0$ (or $\Delta \leq 50$.)

Thus current basis remains optimal iff $100/3 = 50-50/3 \leq$ Amount of Available Sugar $\leq 50 + 50 = 100$.

6d. After this change the current basis is still optimal.

New Profit $= c_{BV}B^{-1}b = [4\ 1] \begin{bmatrix} 60 \\ 100 \end{bmatrix} = \$3.40$

New values of decision variables are found from

$$\begin{bmatrix} x_3 \\ x_2 \end{bmatrix} = B^{-1}b = \begin{bmatrix} 3/2 & -1/2 \\ -1/2 & 1/2 \end{bmatrix} \begin{bmatrix} 60 \\ 100 \end{bmatrix} = \begin{bmatrix} 40 \\ 20 \end{bmatrix}$$

Thus 40 Type 3 Candy Bars, 20 Type 2 Candy Bars, and 0 Type 1 candy bars would now be manufactured. If only 30 ounces of sugar were available the current basis would no longer be optimal and we would have to resolve the problem to find the new optimal solution.

6e. Coefficient of Type 1 Candy Bar in row 0 is now

$[4\ 1] \begin{bmatrix} 1/2 \\ 1/2 \end{bmatrix} -3 = -.5$. Thus current basis is no longer optimal and the new optimal

solution would manufacture Type 1 Candy Bars.

6f. The coefficient of Type 4 Candy Bars in row 0 will now be

$[4 \ 1] \begin{bmatrix} 3 \\ 4 \end{bmatrix}$ - 17 = -1. Thus x_4 should be entered into the basis and the current basis is

no longer optimal. The new optimal solution will make Type 4 Candy Bars.

Section 6.4

4. Both the fat and calorie constraints are nonbinding. The new fat and calorie constraints are each within their allowable range so the current basis remains optimal.

5. We need to use the 100% Rule. Since

$$\frac{30-15}{30} + \frac{80-60}{50} = .90 \leq 1 \text{ the current basis remains optimal.}$$

6. Use the 100% rule. Since

$$\frac{8-6}{4} + \frac{500-60}{\infty} = .50 \leq 1 \text{ the current basis remains optimal.}$$

Section 6.5

1.   min w = $y_1 + 3y_2 + 4y_3$
        s.t. $-y_1 + y_2 + y_3 \geq 2$
                $y_1 + y_2 - 2y_3 \geq 1$
                $y_1, y_2, y_3 \geq 0$

4.   max z = $6x_1 + 8x_2$
        s.t. $x_1 + x_2 \leq 4$
                $2x_1 - x_2 \leq 2$
                $2x_2 = -1$
                $x_1 \leq 0 \ \ x_2 \text{ u.r.s.}$

Section 6.7

1a. min $w = 100y_1 + 80y_2 + 40y_3$
   s.t. $2y_1 + y_2 + y_3 \geq 3$
           $y_1 + y_2 \geq 2$
           $y_1 \geq 0 \ y_2 \geq 0 \ y_3 \geq 0$

1b. and 1c. $y_1 = 1 \ y_2 = 1 \ y_3 = 0 \ w = 180$. Observe that this solution has a w-value that equals the optimal primal z-value. Since this solution is dual feasible it must be optimal (by Lemma 2) for the dual.

4. The dual is
$$\max z = 4x_1 + 20x_2 + 10x_3$$
   s.t. $x_1/2 + x_2 + x_3 \geq 2$
           $x_1/4 + 3x_2 + x_3 \geq 3$
           $x_1, x_2, x_3 \geq 0$
The optimal solution to the dual is $x_1 = 0$, $x_2 = 1/2$, $x_3 = 3/2$, $z = 4(0) + 20(1/2) + 10(3/2) = 25$ cents.

Section 6.8

2a. Shadow Price for Sugar Const. $= 4$ Shadow Price for Chocolate Const. $= 1$. If 1 extra ounce of sugar were available profits would increase by \$4. If 1 extra ounce of chocolate were available profits would increase by \$1. Without further information however we cannot determine how much we would pay for an additional ounce of chocolate or sugar.

2b. From Section 6.3 current basis remains optimal if $100/3 \leq$ Available Sugar ≤ 100.
Here $\Delta b_1 = 60-50 = 10$ so
New z-value $=$ [old z-value] $+ 10(4) = 340$ cents

2c. $\Delta b_1 = -10$ Thus [New z-value] $= 300-10 \ (4) = 260$ cents.

2d. Since current basis is no longer optimal we cannot answer this question without resolving the problem.

Section 6.9

1. Dual constraint for computer tables is $6y_1 + 2y_2 + y_3 \geq 35$. Since [0 10 10] does not satisfy this constraint the current basis is no longer optimal. Another way of seeing it: a computer table uses $20 worth of finishing time and $10 worth of carpentry time. Since a computer table sells for $35 it pays to make computer tables and the current basis is no longer optimal.

Section 6.10

1a. min $w = 600y_1 + 400y_2 + 500y_3$

s.t. $4y_1 + \ y_2 + \ 3y_3 \geq 6$ (1)

      $9y_1 + \ y_2 + \ 4y_3 \geq 10$ (2)

      $7y_1 + 3y_2 + \ 2y_3 \geq 9$ (3)

      $10y_1 + 40y_2 + \ \ y_3 \geq 20$ (4)

      $y_1, y_2, y_3, y_4 \geq 0$

1b. Since $s_3 > 0$, $y_3 = 0$. Since $x_1 > 0$, (1) is binding. Since $x_4 > 0$, (4) is binding. Setting $y_3 = 0$ and solving (1) and (4) simultaneously yields $y_1 = 22/15$ $y_2 = 2/15$. Thus the optimal dual solution is $y_1 = 22/15$, $y_2 = 2/15$, $y_3 = 0$, $z = 2800/3$.

1c. (40): $s_3 > 0$ implies $y_3 = 0$. This is reasonable because if all available glass is not being used an additional ounce of glass will not increase z hence glass constraint should have 0 shadow price.

   (41): $y_2 > 0$ implies $s_2 = 0$. Since $y_2 > 0$, an additional minute of packaging time will increase z. Thus all packaging time that is currently available must be used (hence $s_2 = 0$).

   (42): $e_2 = 9(22/15) + 2/15 - 10 = 50/15 > 0$ implies $x_2 = 0$. Note that $e_2 =$ [Cost of Resources Used to Make a Beer glass] -[Price of a Beer Glass]. Since $e_2 > 0$, producing a beer glass would not be profitable so $x_2 = 0$ should hold.

   (43): $x_1 > 0$ implies $e_1 = 0$. Since $x_1 > 0$, we are manufacturing wine glasses. Thus MR = MC yields that Sale Price of a Wine Glass = Cost of Making a Wine Glass or [Cost of Making a Wine Glass]-[Sales Price of Wine Glass] = 0 Thus $e_1 = 4(22/15) + 1(2/15) + 3(0)-6 = 0$.

Section 6.11

3. The rhs of the optimal tableau is now

$$\begin{bmatrix} 1 & 2 & -8 \\ 0 & 2 & -4 \\ 0 & -.5 & 1.5 \end{bmatrix} \begin{bmatrix} 20 \\ 20 \\ 8 \end{bmatrix} = \begin{bmatrix} -4 \\ 8 \\ 2 \end{bmatrix}$$

rhs of row 0 is $[0\ 10\ 10]$ $\begin{bmatrix} 20 \\ 20 \\ 8 \end{bmatrix} = 280$

We can apply the dual simplex to the following tableau:

z	x_1	x_2	x_3	s_1	s_2	s_3	RHS
1	0	5	0	0	10	10	280
0	0	-2	0	1	2	-8	-4
0	0	-2	1	0	2	-4	8
0	1	1.25	0	0	-.5	1.5	2
Ratio		2.5				1.25	

s_3 now enters the basis in row 1. This pivot yields an optimal tableau with $z = 275$ s_3 $= 1/2$ $x_3 = 10$ $x_1 = 1.25$.

Section 6.12

1. From printouts we find that only School 4 is inefficient. We find that .065(School 2 Inputs) + 1.21(School 3 Inputs) =

$$\begin{bmatrix} 14.22 \\ 7.59 \\ .08 \end{bmatrix}$$

Thus composite school uses 5% less of Inputs 1 and 2.

$$.065(\text{School 2 Outputs}) + 1.21(\text{School 3 Outputs}) = \begin{bmatrix} 13.96 \\ 9 \\ 10.13 \end{bmatrix}$$

Thus composite school can produce 55% better reading with fewer inputs.

Problem 1 School 4 Printout

```
MAX 9T1+9T2+9T3
ST
-9T1-7T2-6T3+13W1+4W2+.05W3>0
-10T1-8T2-7T3+14W1+5W2+.05W3>0
-11T1-7T2-8T3+11W1+6W2+.06W3>0
-9T1-9T2-9T3+15W1+8W2+.08W3>0
T1>.0001
T2>.0001
T3>.0001
W1>.0001
W2>.0001
W3>.0001
15W1+8W2+.08W3=1
END
```

LP OPTIMUM FOUND AT STEP   6

OBJECTIVE FUNCTION VALUE

1)   0.9484789

VARIABLE	VALUE	REDUCED COST
T1	0.000100	0.000000
T2	0.105187	0.000000
T3	0.000100	0.000000
W1	0.047177	0.000000
W2	0.000100	0.000000
W3	3.644364	0.000000

ROW	SLACK OR SURPLUS	DUAL PRICES
2)	0.058110	0.000000
3)	0.000000	-0.065455
4)	0.000000	-1.210909
5)	0.051521	0.000000
6)	0.000000	-4.974545
7)	0.105087	0.000000
8)	0.000000	-1.145455
9)	0.047077	0.000000
10)	0.000000	0.000000
11)	3.644264	0.000000
12)	0.000000	0.949091

Chapter 7 Solutions

Section 7.1

1.

	Cust. 1	Cust. 2	Cust. 3	
Warehouse 1	15	35	25	40
Warehouse 2	10	50	40	30
Shortage	90	80	110	20
	30	30	30	

2.

	C1	C2	C3	DUMMY	
W1	15	35	25	0	40
W2	10	50	40	0	30
W1 EXTRA	115	135	125	0	20
W2 EXTRA	110	150	140	0	20
	30	30	30	20	

Section 7.2

1. By NW corner method we obtain following bfs for Problem 1:

30	10		40
	20	10	30
		20	20
30	30	30	

By NW corner method we obtain the following bfs for Problem 2:

	C1	C2	C3	DUMMY	
W1	15 (30)	35 (10)	25	0	40
W2	10	50 (20)	40 (10)	0	30
W1 EXTRA	115	135	125 (20)	0 (0)	20
W2 EXTRA	110	150	140	0 (20)	20
	30	30	30	20	

By NW corner method we obtain following bfs for Problem 3:

	M1	M2	M3	M4	M5	M6	DUMMY	
1R	200							200
1O	0	100						100
2R		160	40					200
2O			100					100
3R			100	100				200
3O				100				100
4R				140	60			200
4O					100			100
5R					30	150	20	200
5O							100	100
6R							200	200
6O							100	100
	200	260	240	340	190	150	420	

Section 7.3

1. We begin with the bfs obtained in Section 7.2

	v's → 15	35	25	
u's ↓				
0	15 (30)	35 (10)	25	40
15	10	50 (20)	40 (10)	30
85	90	80	110 (20)	20
	30	30	30	

Since $\bar{c}_{32}=40$ we enter x_{32} into basis. The loop involving x_{32} and some of the basic variables is $(3, 2) - (2, 2) - (2, 3) - (3, 3)$. x_{33} exits yielding the following bfs

40

u's \ v's	15	35	25	
0	15 30	35 10	25	40
15	10 0	50 30	40	30
45	90	80 20	110	20
	30	30	30	

Since $\bar{c}_{21}=20$ we enter x_{21} into the basis. The loop involving x_{21} and some of the basic variables is (2,1)-(1,1)-(1,2)-(2,2). After x_{22} exits we obtain the following bfs:

u's \ v's	15	35	45	
0	15 30	35 10	25	40
-5	10 0	50	40 30	30
45	90	80 20	110	20
	30	30	30	

Now $\bar{c}_{13}=20$ so we enter x_{13}. The relevant loop is (1, 3) - (2, 3) - (2, 1) - (1, 1). After x_{11} exits we obtain the following bfs:

u's \ v's	-5	35	25	
0	15	35 10	25 30	40
15	10 30	50	40 0	30
45	90	80 20	110	20
	30	30	30	

This is an optimal tableau. Thus 10 units should be sent from Warehouse 1 to Customer 2, 30 units from Warehouse 1 to Customer 3, 30 units from Warehouse 2 to customer 1. 20 units of Customer 2's demand will be unsatisfied.

2.

	C1	C2	C3	DUMMY	
W1	15 30	35 10	25	0	40
W2	10	50 20	40 10	0	30
W1 EXTRA	115	135 20	125 0	0	20
W2 EXTRA	110	150	140	0 20	20
	30	30	30	20	

x_{21} enters and x_{22} exits yielding

	C1	C2	C3	DUMMY	
W1	15 10	35 30	25	0	40
W2	10 20	50	40 10	0	30
W1 EXTRA	115	135 20	125 0	0	20
W2 EXTRA	110	150	140	0 20	20
	30	30	30	20	

x_{13} enters and either x_{11} or x_{23} exits (we choose x_{23} to exit) yielding

	C1	C2	C3	DUMMY	
W1	15 0	35 30	25 10	0	40
W2	10 30	50	40	0	30
W1 EXTRA	115	135 20	125 0	0	20
W2 EXTRA	110	150	140	0 20	20
	30	30	30	20	

x_{41} enters yielding the following optimal tableau:

	C1	C2	C3	DUMMY	
	15	35	25	0	
W1		30	10		40
	10	50	40	0	
W2	30				30
	115	135	125	0	
W1 EXTRA		20	0		20
	110	150	140	0	
W2 EXTRA	0			20	20
	30	30	30	20	

Section 7.4

1. x_{14} is a non-basic variable in the optimal solution. If we change c_{14} to $9 + \Delta$, we find that

$$\bar{c}_{14} = 0 + 2 - (9 + \Delta) = -7 - \Delta.$$

Thus the current basis remains optimal for $-7 - \Delta \leq 0$ or $\Delta \geq -7$. Thus the current basis remains optimal for $c_{14} \geq 9 - 7 = 2$.

3. x_{23} is a basic variable in the optimal tableau. Thus the new optimal solution leaves all variables the same except for x_{23}, which is increased by 3 to $x_{23} = 5 + 3 = 8$. z increases by $3(13)$, so the new optimal z-value is $1020 + 39 = 1059$.

Section 7.5

1. We use a cost of M to rule out forbidden assignments and add a dummy job (Job 5) to balance the problem. Assigning a person to the dummy job has a 0 cost.

	Job						Row Min
		1	2	3	4	5	
	1	22	18	30	18	0	0
Person	2	18	M	27	22	0	0
	3	26	20	28	28	0	0
	4	16	22	M	14	0	0
	5	21	M	25	28	0	0
Column Min		16	18	25 14		0	

43

Since all Row minima are 0 we proceed to the column minima. The reduced cost matrix is

Job

		1	2	3	4	5
	1	6	0	5	4	0
	2	2	M	2	8	0
Person	3	10	2	3	14	0
	4	0	4	M	0	0
	5	5	M	0	14	0

Only 4 lines (we have used Row 1, Row 4, Col. 3 and Col. 5) are needed to cover all 0's in this matrix. The smallest uncovered element is 2 so we subtract 2 from all uncovered costs and add 2 to all twice covered costs. The resulting matrix is

Job

		1	2	3	4	5
	1	6	0	7	4	2
	2	0	M	2	6	0
Person	3	8	0	3	12	0
	4	0	4	M	0	2
	5	3	M	0	12	0

5 lines are needed to cover the 0's so an optimal solution is available: $x_{12}=1$, $x_{21}=1$, $x_{35}=1$, $x_{44}=1$, $x_{53}=1$. Thus Person 3 is not assigned a job and a total time of $18+18+14+25=75$ time units is required to complete all the jobs.

4. Cost matrix and optimal assignments (denoted by *) are as follows:

Job

Person	1	2	3	4	Dummy
1	50	46	42	40*	0
2	51*	48	44	1000	0
2'	51	48	44*	1000	0
3	1000	47	45	45	0
3'	1000	47	45	45	0

Note: 1000 rules out prohibited assignment. Total cost $= 182$.

Section 7.6

2. (Supplies and demands are in thousands) Total Supply-Total Demand=350-300=50 so dummy demand pt. has demand of 50. We obtain the following balanced transportation tableau:

	Mobile	Galv.	NY	LA	Dummy	Supplies
Well 1	10	13	25	28	0	150
Well 2	15	12	26	25	0	200
Mobile	0	6	16	17	0	0+350=350
Galv.	6	0	14	16	0	0+350=350
NY	M	M	0	15	0	0+350=350
LA	M	M	15	0	0	0+350
Demands	350	350	140 +350	160 +350	50	

45

Chapter 8 Solutions

Section 8.2

1. First label node 1 with a permanent label: [0* 7 12 21 31 44]
   Now node 2 receives a permanent label [0* 7* 12 21 31 44].

Node   Temporary Label (* denotes next assigned permanent label)
3      $\min\{12,7+7\} = 12^*$
4      $\min\{21,7+12\} = 19$
5      $\min\{31,7+21\} = 28$
6      $\min\{44,7+31\} = 38$
Now labels are [0* 7* 12* 19 28 38]

Node   Temporary Label (* denotes next assigned permanent label)
4      $\min\{19,12+7\} = 19^*$
5      $\min\{28,12+12\} = 24$
6      $\min\{38,12+21\} = 33$
Now labels are [0* 7* 12* 19* 24 33]

Node   Temporary Label (* denotes next assigned permanent label)
5      $\min\{24,19+7\} = 24^*$
6      $\min\{33, 19+12\} = 31$
Now labels are [0* 7* 12* 19* 24* 31]
Node   Temporary Label (* denotes next assigned permanent label)
6      $\min\{31,24+7\} = 31$
Now labels are [0* 7* 12* 19* 24* 31*]
$31 - 24 = c_{56}$, $24 - 12 = c_{35}$, $12 - 0 = c_{13}$. Thus 1-3-6 is the shortest path (of length 31) from node 1 to node 6.

Section 8.3

1. $\max z = x_0$
        s.t. $x_{so,1} \leq 6$, $x_{so,2} \leq 2$ $x_{12} \leq 1$, $x_{32} \leq 3$, $x_{13} \leq 3$, $x_{3,si} \leq 2$, $x_{24} \leq 7$, $x_{4,si} \leq 7$
$x_0 = x_{so,1} + x_{so,2}$ (Node so)
$x_{so,1} = x_{13} + x_{12}$ (Node 1)
$x_{12} + x_{32} + x_{s0,2} = x_{24}$ (Node 2)

$x_{13} = x_{32} + x_{3,si}$ (Node 3)

$x_{24} = x_{4,si}$ (Node 4)

$x_{3,si} + x_{4,si} = x_0$ (Node si)

All variables ≥ 0

Initial flow of 0 in each arc. Begin by labeling sink via path of forward arcs (so, 1) - (1, 3) - (3, 2) - (2, 4) - (4, si). Increase flow in each of these feasible arcs by 3, yielding the following feasible flow:

Arc	Flow
so-1	3
so-2	0
1-3	3
1-2	0
2-4	3
3-si	0
3-2	3
4-si	3
Flow to sink 3	

Now label sink by (so-2) - (2-4), (4, si). Each arc is a forward arc and we can increase flow in each arc by 2. This yields the following feasible flow:

Arc	Flow
so-1	3
so-2	2
1-3	3
1-2	0
2-4	5
3-si	0
3-2	3
4-si	5
Flow to sink 5	

Now label sink by (so-1) - (1, 2) - (3, 2) - (3, si). All arcs on this path are forward arcs except for (3, 2), which is a backwards arc. We can increase the flow on each forward arc by 1 and decrease the flow on each backward arc by 1. This yields the following feasible flow:

Arc	Flow
so-1	4
so-2	2
1-3	3

48

1-2	1
2-4	5
3-si	1
3-2	2
4-si	5
Flow to sink 6	

The sink cannot be labeled, so we found the maximum flow of 6 units. The minimum cut is obtained from V' = {3, 2, 4, si}. This cut consists of arcs (1, 3), (1, 2), (so, 2) and has capacity of 3 + 1 + 2 = 6 = maximum flow.

4.   Maximum flow is 45. Min Cut Set = {1, 3, and si}. Capacity of Cut Set = 20 + 15 + 10 = 45. See Figure.

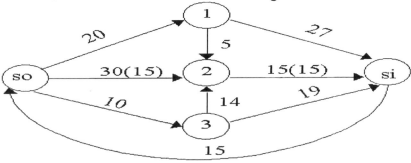

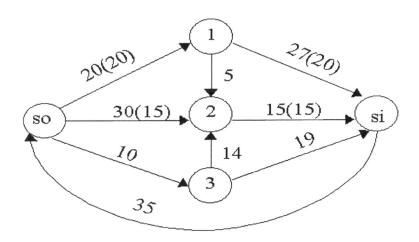

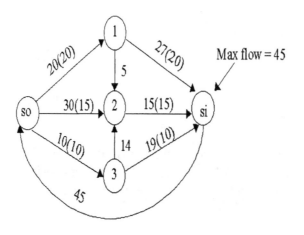

Max flow = 45

Section 8.4

2.	Activity	Predecessors	Duration (wks.)
	A = Design	-	5
	B = Make Part A	A	4
	C = Make Part B	A	5
	D = Make Part C	A	3
	E = Test Part A	B	2
	F = Assemble A and B	C,E	2
	G = Attach C	D,F	1
	H = Test Final Product	G	1

From the project diagram we find that

ET(1) = 0	LT(1) = 0	TF(1,2) = 0	FF(1,2) = 0
ET(2) = 5	LT(2) = 5	TF(2,3) = 0	FF(2,3) = 0
ET(3) = 9	LT(3) = 9	TF(3,4) = 0	FF(3,4) = 0
ET(4) = 11	LT(4) = 11	TF(2,4) = 1	FF(2,4) = 1
ET(5) = 13	LT(5) = 13	TF(4,5) = 0	FF(4,5) = 0
ET(6) = 14	LT(6) = 14	TF(2,5) = 5	FF(2,5) = 5
ET(7) = 15	LT(7) = 15	TF(5,6) = 0	FF(5,6) = 0
		TF(6,7) = 0	FF(6,7) = 0

Looking at the activities with TF of 0, we find that the critical path is 1-2-3-4-5-6-7 (length 15 days).

The appropriate LP is

$$\min z = x_7 - x_1$$

s.t.
$$x_2 \geq x_1 + 5$$
$$x_3 \geq x_2 + 4$$
$$x_4 \geq x_3 + 2$$
$$x_4 \geq x_2 + 5$$
$$x_5 \geq x_2 + 3$$
$$x_5 \geq x_4 + 2$$
$$x_6 \geq x_5 + 1$$
$$x_7 \geq x_6 + 1$$

all variables urs

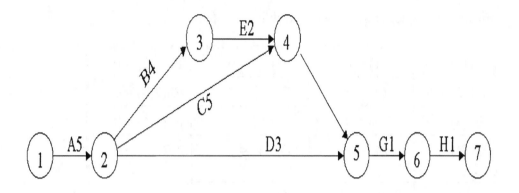

5a.From the project diagram we find that

ET(1) = 0	LT(1) = 0	TF(1,2) = 0	FF(1,2) = 0
ET(2) = 5	LT(2) = 5	TF(2,3) = 0	FF(2,3) = 0
ET(3) = 13	LT(3) = 13	TF(3,5) = 0	FF(3,5) = 0
ET(4) = 17	LT(4) = 17	TF(3,6) = 8	FF(3,6) = 8
ET(5) = 23	LT(5) = 23	TF(3,4) = 0	FF(3,4) = 0
ET(6) = 26	LT(6) = 26	TF(4,5) = 0	FF()4,5 = 0

Both are 1-2-3-5-6 and 1-2-3-4-5-6 are critical paths having length 26 days.

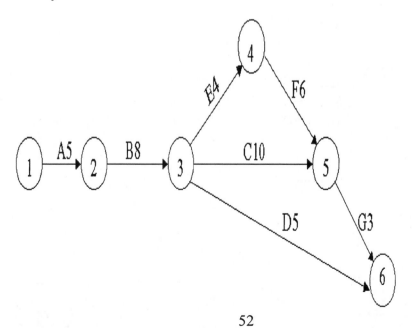

5b. Let A = number of days by which we reduce duration of activity A, etc. and x_j = time that event at node j occurs

min z = 30A + 15B + 20C + 40D + 20E + 30F + 40G

$A \leq 2$, $B \leq 3$, $C \leq 1$, $D \leq 2$, $E \leq 2$, $F \leq 3$, $G \leq 1$

$x_2 \geq x_1 + 5 - A$

$x_3 \geq x_2 + 8 - B$

$x_4 \geq x_3 + 4 - E$

$x_5 \geq x_3 + 10 - C$

$x_5 \geq x_4 + 6 - F$

$x_6 \geq x_5 + 3 - G$

$x_6 \geq x_3 + 5 - D$

$x_6 - x_1 \leq 20$

A,B,C,D,E,F,G ≥ 0, x_j urs

Section 8.5

1. min z = $4x_{12} + 3x_{24} + 2x_{46} + 3x_{13} + 3x_{35} + 2x_{25} + 2x_{56}$

st $x_{12} + x_{13} = 1$ (node 1 constraint)

$x_{12} = x_{24} + x_{25}$ (node 2 constraint)

$x_{13} = x_{35}$ (node 3 constraint)

$x_{24} = x_{46}$ (node 4 constraint)

$x_{25} = x_{56}$ (node 5 constraint)

$x_{46} + x_{56} = 1$ (node 6 constraint)

all $x_{ij} \geq 0$

If $x_{ij} = 1$ the shortest path from node 1 to node 6 will contain arc (i, j) while if $x_{ij} = 0$ the shortest path from node 1 to node 6 does not contain arc (i, j).

3. Node   Net Outflow

Detroit   6500

Dallas   6000

City 1   -5000

City 2   -4000

City 3   -3000

Dummy   -500

All arcs from Detroit or Dallas to City 1, 2, or 3 have a capacity of 2200. Other arcs have infinite capacity

Arc	Shipping Cost
Detroit-City 1	$2800
Detroit-City 2	$2600
Detroit-City 3	$2300
Detroit-Dummy	$0
Dallas-City 1	$2300
Dallas-City 2	$2000
Dallas-City 3	$2000
Dallas-Dummy	$0

Section 8.6

1. We begin at Gary and include the Gary-South Bend arc. Then we add the South Bend-Fort Wayne arc. Next we add the Gary-Terre Haute arc. Finally we add the Terre Haute-Evansville arc. This minimum spanning tree has a total length of $58 + 79 + 164 + 113 = 414$ miles.

Section 8.7

3.) Find the Y values.

$Y_1 = 0$, $Y_1 - Y_2 = 15$, $Y_2 - Y_4 = 5$, $Y_2 - Y_5 = 10$, $Y_3 - Y_4 = 4$

This yields

$Y_1 = 0$, $Y_2 = -15$, $Y_3 = -16$, $Y_4 = -20$, $Y_5 = -25$

Find the row 0 coefficients for each non-basic variable.

$\overline{C}_{13} = 0-(-16)-11 = 5$ (satisfies optimality condition)

$\overline{C}_{23} = -15-(-16)-5 = -4$ (satisfies optimality condition)

$\overline{C}_{35} = -16-(-25)-5 = 4$ (violates optimality condition)

$\overline{C}_{45} = -20-(-25)-14 = -9$ (satisfies optimality condition)

Enter X_{35} into the basis.

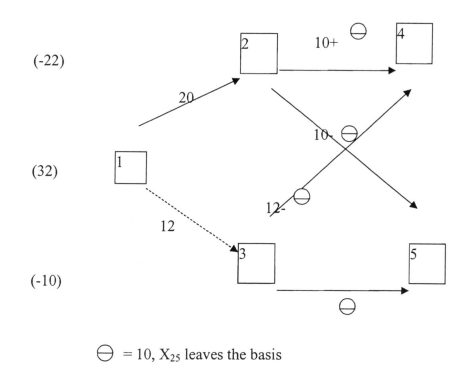

(-22)

(32)

(-10)

$\ominus$ = 10, X_{25} leaves the basis

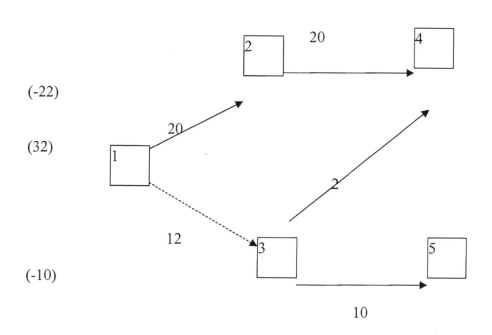

(-22)

(32)

(-10)

Find the Y values.

$Y_1 = 0$, $Y_1 - Y_2 = 15$, $Y_2 - Y_4 = 5$, $Y_3 - Y_4 = 4$, $Y_3 - Y_5 = 5$

This yields

$Y_1 = 0$, $Y_2 = -15$, $Y_3 = -16$, $Y_4 = -20$, $Y_5 = -21$

Find the row 0 coefficients for each non-basic variable.

$\overline{C}_{13} = 0-(-16)-11 = 5$ (satisfies optimality condition)

$\overline{C}_{23} = -15-(-16)-5 = -4$ (satisfies optimality condition)

$\overline{C}_{25} = -15-(-21)-10 = -4$ (satisfies optimality condition)

$\overline{C}_{45} = -20-(-21)-14 = -13$ (satisfies optimality condition)

Thus, an optimal solution to this MCNFP is
(Basic Variables) $X_{12} = 20$, $X_{24} = 20$, $X_{34} = 2$, $X_{35} = 10$
(Non-basic Variables at upper bound) $X_{13} = 12$
(Non-basic Variables at lower bound) $X_{23} = 0$, $X_{25} = 0$, $X_{45} = 0$

The optimal z-value is
$Z = 20(15) + 12(11) + 20(5) + 2(4) + 10(5) = \$590$

Chapter 9 Solutions

Section 9.2

1. 1. Let $x_i = 1$ if player i starts

$x_i = 0$ otherwise

Then appropriate IP is

$$\max z = 3x_1 + 2x_2 + 2x_3 + x_4 + 3x_5 + 3x_6 + x_7$$

s.t.　$x_1 + x_3 + x_5 + x_7 \geq 4$ (guards)

$x_3 + x_4 + x_5 + x_6 + x_7 \geq 2$ (forwards)

$x_2 + x_4 + x_6 \geq 1$ (center)

$x_1 + x_2 + x_3 + x_4 + x_5 + x_6 + x_7 = 5$

$3x_1 + 2x_2 + 2x_3 + x_4 + 3x_5 + 3x_6 + 3x_7 \geq 10$ (BH)

$3x_1 + x_2 + 3x_3 + 3x_4 + 3x_5 + x_6 + 2x_7 \geq 10$ (SH)

$x_1 + 3x_2 + 2x_3 + 3x_4 + 3x_5 + 2x_6 + 2x_7 \geq 10$ (REB)

$x_6 + x_3 \leq 1$

$-x_4 - x_5 + 2 \leq 2y$

(If $x_1 > 0$ then $x_4 + x_5 \geq 2$)

$x_1 \quad \leq 2(1-y)$

$x_2 + x_3 \geq 1$

$x_1, x_2, \ldots x_7, y$ are all 0-1 variables

3. Let x_1 = Units of Product 1 produced

x_2 = Units of Product 2 produced

$y_i = 1$ if any Product i is produced

$y_i = 0$ otherwise

Then the appropriate IP is

$$\max z = 2x_1 + 5x_2 - 10y_1 - 20y_2$$

s.t.　$3x_1 + 6x_2 \leq 120$

$x_1 \quad \leq 40y_1$

$x_2 \leq 20y_2$

$x_1 \geq 0, x_2 \geq 0, \ y_1, y_2 = 0$ or 1

14a. Let $x_i = 1$ if disk i is used, $x_i = 0$ otherwise

$$\min z = 3x_1 + 5x_2 + x_3 + 2x_4 + x_5 + 4x_6 + 3x_7 + x_8 + 2x_9 + 2x_{10}$$

s.t.　$x_1 + x_2 + x_4 + x_5 + x_8 + x_9 \geq 1$ (File 1)

$x_1 + x_3 \geq 1$ 　(File 2)

$x_2 + x_5 + x_7 + x_{10} \geq 1$ (File 3)

$x_3 + x_6 + x_8 \geq 1$ (File 4)

$x_1 + x_2 + x_4 + x_6 + x_7 + x_9 + x_{10} \geq 1$ (File 5)

$x_i = 0$ or 1 (i=1,2,...10)

14b. If $x_3 + x_5 > 0$, then $x_2 \geq 1$ yields

$1 - x_2 \leq 2y$

$x_3 + x_5 \leq 2(1 - y)$ y=0 or 1

(need M=2 because $x_3 + x_5 = 2$ is possible)

Section 9.3

1.

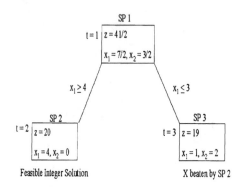

SP 1

t = 1 | z = 41/2

$x_1 = 7/2, x_2 = 3/2$

$x_1 \geq 4$ $x_1 \leq 3$

SP 2   SP 3

t = 2 | z = 20   t = 3 | z = 19

$x_1 = 4, x_2 = 0$ $x_1 = 1, x_2 = 2$

Feasible Integer Solution   X beaten by SP 2

Optimal Solution is z = 20, $x_1 = 4, x_2 = 0$.

2. We wish to solve min $z = 50x_1 + 100x_2$

st $7x_1 + 2x_2 \geq 28$

$2x_1 + 12x_2 \geq 24$

$x_1, x_2 \geq 0$

SP 1

st $7x_1 + 2x_2 \geq 28$

$2x_1 + 12x_2 \geq 24$

$x_1, x_2 \geq 0$

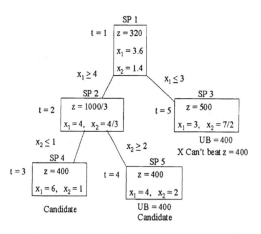

SP 1

t = 1 | z = 320

$x_1 = 3.6$

$x_1 \geq 4$ $x_2 = 1.4$ $x_1 \leq 3$

SP 2   SP 3

t = 2 | z = 1000/3   t = 5 | z = 500

$x_1 = 4, x_2 = 4/3$ $x_1 = 3, x_2 = 7/2$

UB = 400

$x_2 \leq 1$ $x_2 \geq 2$ X Can't beat z = 400

SP 4   SP 5

t = 3 | z = 400   t = 4 | z = 400

$x_1 = 6, x_2 = 1$ $x_1 = 4, x_2 = 2$

Candidate   UB = 400

Candidate

58

Note: In solving subproblems we have used the result discussed in Problem 8 to conclude that in each subproblem's optimal solution the "newest" constraint must be binding. Thus we know that some optimal solution to SP2 will have $x_1 = 4$.

The two optimal solutions $x_1 = 6$, $x_2 = 1$ and $x_1 = 4$, $x_2 = 2$ have been found.

Section 9.4

3.

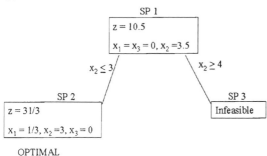

OPTIMAL

Section 9.5

3.

Letting $x_i = 1$ if item i is chosen and $x_i = 0$ otherwise yields the following knapsack problem:

$$\max z = 5x_1 + 8x_2 + 3x_3 + 7x_4$$
$$\text{st } 3x_1 + 5x_2 + 2x_3 + 4x_4 \le 6$$
$$x_i = 0 \text{ or } 1$$

We obtain the following tree (for each subproblem any omitted variable equals 0):

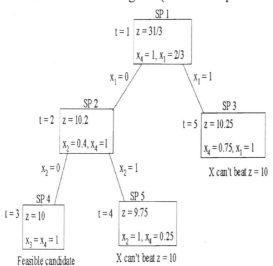

59

Note that since optimal objective function value for any candidate solution associated with a branch must be an integer, SP 3 can at best yield a z-value of 10, so we need not branch on SP 3. Thus the optimal solution is $z = 10$, $x_1 = x_2 = 0$, $x_3 = x_4 = 1$.

Section 9.6

Let LFR = City 1, LFP = City 2, LR = City 3, and LP = City 4. Then we obtain the following branch and bound tree:

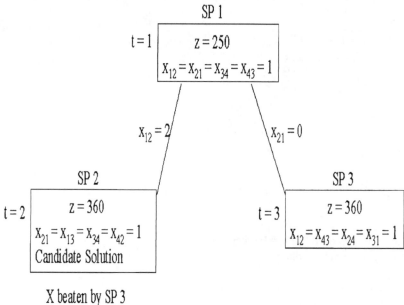

1-2-4-3-1 with total setup time of 330 minutes is optimal. Thus we should produce gasolines in order LFR-LFP-LP-LR-LFR.

3. Add a city 1' and arcs (2,1'), (3,1'), and (4,1'). Cost of arc from (i,1') = cost of arc from (i,1). Now find shortest Hamiltonian path from 1 to 1'.

Section 9.7

3. Let $x_i = 1$ if project i is chosen and $x_i = 0$ otherwise. Then the appropriate IP is:

$$\max z = 5x_1 + 9x_2 + 6x_3 + 3x_4 + 2x_5$$
$$\text{st } 4x_1 + 6x_2 + 5x_3 + 4x_4 + 3x_5 \leq 10$$
$$x_1 + x_2 \leq 1, \ x_3 + x_4 \leq 1, \ x_2 \leq x_5$$
$$\text{All } x_i = 0 \text{ or } 1$$

60

From the tree we find an optimal solution is $z = 11$,
$x_1 = x_3 = 1$, $x_2 = x_4 = x_5 = 0$.

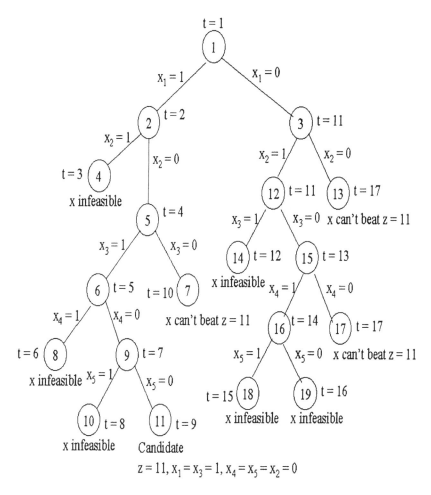

Candidate

$z = 11$, $x_1 = x_3 = 1$, $x_4 = x_5 = x_2 = 0$

Section 9.8

1. Since both constraints have a fractional part of 1/2 in the optimal tableau, we arbitrarily choose to use the first constraint to yield the cut:

$x_2 + 7s_1/22 + s_2/22 = 3 + 1/2$ or

$x_2 - 3 = 1/2 - 7s_1/22 - s_2/22$ or

$1/2 - 7s_1/22 - s_2/22 \leq 0$. Adding this constraint with a slack variable s_3
yields the following tableau:

z	x_1	x_2	s_1	s_2	s_3	RHS
1	0	0	56/11	30/11	0	126
0	0	1	7/22	1/22	0	7/2
0	1	0	-1/22	3/22	0	9/2
0	0	0	-7/22	-1/22	1	-1/2

The dual simplex pivoting rule indicates that s_1 should enter in row 3 yielding the following tableau:

z	x_1	x_2	s_1	s_2	s_3	RHS
1	0	0	0	2	16	118
0	0	1	0	0	1	3
0	1	0	0	1/7	-1/7	32/7
0	0	0	1	1/7	-22/7	11/7

We now arbitrarily choose row 2 to generate the next cut:
$x_1 + s_2/7 - s_3 + 6s_3/7 = 4/7 + 4$ or
$x_1 - s_3 - 4 = 4/7 - s_2/7 - 6s_3/7$ yielding the cut
$- s_2/7 - 6s_3/7 \leq -4/7$. Adding a slack variable s_4 to the cut yields the following tableau:

z	x_1	x_2	s_1	s_2	s_3	s_4	RHS
1	0	0	0	2	16	0	118
0	0	1	0	0	1	0	3
0	1	0	0	1/7	-1/7	0	32/7
0	0	0	1	1/7	-22/7	0	11/7
0	0	0	0	-1/7	-6/7	1	-4/7

Entering s_2 in the last constraint yields the following (optimal) tableau.

z	x_1	x_2	s_1	s_2	s_3	s_4	RHS
1	0	0	0	0	4	14	110
0	0	1	0	0	1	0	3
0	1	0	0	0	-1	1	4
0	0	0	1	0	-4	1	1
0	0	0	0	1	6	-7	4

This tableau yields the optimal solution z = 110 x_1 = 4 x_2 = 3.

Chapter 10 Solutions

Section 10.1

$\max z = 3x_1 + x_2 + x_3$

s.t. $x_1 + x_2 + x_3 + s_1 = 6$

$2x_1 - x_3 + s_2 = 4$

$x_2 + x_3 + s_3 = 2$ $\qquad x_1, x_2, x_3 \geq 0$

$BV(0) = \{s_1, s_2, s_3\}$ $\qquad NBV(0) = \{x_1, x_2, x_3\}$

$$B_0^{-1} = B_0 = \begin{bmatrix} 1 & 0 & 0 \\ 0 & 1 & 0 \\ 0 & 0 & 1 \end{bmatrix} \qquad c_{BV} = [0 \ 0 \ 0]$$

$$c_{BV} B_0^{-1} = [0 \ 0 \ 0]$$

$$\overline{c_1} = [0 \ 0 \ 0] \begin{bmatrix} 1 \\ 2 \\ 0 \end{bmatrix} -3 = -3$$

$$\overline{c_2} = [0 \ 0 \ 0] \begin{bmatrix} 1 \\ 0 \\ 1 \end{bmatrix} \quad -1 = -1$$

$$\overline{c_3} \equiv [0 \ 0 \ 0] \begin{bmatrix} 1 \\ -1 \\ 1 \end{bmatrix} -1 = -1$$

$$\text{Column for } x_1 = \begin{bmatrix} 1 & 0 & 0 \\ 0 & 1 & 0 \\ 0 & 0 & 1 \end{bmatrix} \begin{bmatrix} 1 \\ 2 \\ 0 \end{bmatrix} = \begin{bmatrix} 1 \\ 2 \\ 0 \end{bmatrix}$$

$$\text{Right Hand Side} = \begin{bmatrix} 1 & 0 & 0 \\ 0 & 1 & 0 \\ 0 & 0 & 1 \end{bmatrix} \begin{bmatrix} 6 \\ 4 \\ 2 \end{bmatrix} = \begin{bmatrix} 6 \\ 4 \\ 2 \end{bmatrix}$$

$$BV(1) = \{s_1, x_1, s_3\} \qquad NBV(1) = \{x_2, x_3, s_2\}$$

$$B_1^{-1} = \begin{bmatrix} 1 & -1/2 & 0 \\ 0 & 1/2 & 0 \\ 0 & 0 & 1 \end{bmatrix} \qquad c_{BV}B_1^{-1} = [0 \ 3 \ 0] \begin{bmatrix} 0 & -1/2 & 0 \\ 0 & 1/2 & 0 \\ 0 & 0 & 1 \end{bmatrix} = [0 \ 3/2 \ 0]$$

$$\overline{c_2} = [0 \ 3/2 \ 0] \begin{bmatrix} 1 \\ 0 \\ 1 \end{bmatrix} -1 = -1 \text{ cbar for } s_2 = 3/2\overline{}$$

$$\overline{c_3} = [0 \ 3/2 \ 0] \begin{bmatrix} 1 \\ -1 \\ 1 \end{bmatrix} -1 = -5/2$$

$$\text{Column for } x_3 = \begin{bmatrix} 1 & -1/2 & 0 \\ 0 & 1/2 & 0 \\ 0 & 0 & 1 \end{bmatrix} \begin{bmatrix} 1 \\ -1 \\ 1 \end{bmatrix} = \begin{bmatrix} 3/2 \\ -1/2 \\ 1 \end{bmatrix}$$

$$\text{Right Hand side tableau 1} = \begin{bmatrix} 1 & -1/2 & 0 \\ 0 & 1/2 & 0 \\ 0 & 0 & 1 \end{bmatrix} \begin{bmatrix} 6 \\ 4 \\ 2 \end{bmatrix} = \begin{bmatrix} 4 \\ 2 \\ 2 \end{bmatrix}$$

$$BV(2) = \{s_1, x_1, x_3\} \qquad NBV(2) = \{x_2, s_2, s_3\}$$

$$B_2^{-1} = \begin{bmatrix} 1 & -1/2 & -3/2 \\ 0 & 1/2 & 1/2 \\ 0 & 0 & 1 \end{bmatrix} \quad c_{BV} B_2^{-1} = [0 \ 3 \ 1] \begin{bmatrix} 1 & -1/2 & -3/2 \\ 0 & 1/2 & 1/2 \\ 0 & 0 & 1 \end{bmatrix}$$

$$=[0 \ 3/2 \ 5/2]$$

$$\overline{c_2} = [0 \ 3/2 \ 5/2] \begin{bmatrix} 1 \\ 0 \\ 1 \end{bmatrix} - 1 = 3/2, \text{ cbarfor } s_2 > 0, \text{ cbar for } s_3 > 0$$

so this an optimal tableau.

$$\text{Right Hand Side} = \begin{bmatrix} 1 & -1/2 & -3/2 \\ 0 & 1/2 & 1/2 \\ 0 & 0 & 1 \end{bmatrix} \begin{bmatrix} 6 \\ 4 \\ 2 \end{bmatrix} = \begin{bmatrix} 1 \\ 3 \\ 2 \end{bmatrix}$$

$$\begin{bmatrix} S_1 \\ X_1 \\ X_3 \end{bmatrix} = \begin{bmatrix} 1 \\ 3 \\ 2 \end{bmatrix}$$

$$z = c_{BV} \, B_2^{-1} \, b = \begin{bmatrix} 0 & 3/2 & 5/2 \end{bmatrix} \begin{bmatrix} 6 \\ 4 \\ 2 \end{bmatrix} = 11$$

Section 10.2

1.
In tableau 0, x_1 enters in row 2 .: r = 2, k = 1

$$\begin{bmatrix} \overline{a_{11}} \\ \underline{a_{21}} \\ a_{31} \end{bmatrix} = \begin{bmatrix} 1 \\ 2 \\ 0 \end{bmatrix}$$

$$E_0 = \begin{bmatrix} 1 & -1/2 & 0 \\ 0 & 1/2 & 0 \\ 0 & 0 & 1 \end{bmatrix} = E_0 \, B_0^{-1} = B_1^{-1}$$

In tableau 1, x_3 enters in row 3: r = 3, k = 3

$$\begin{bmatrix} \overline{a_{13}} \\ \underline{a_{23}} \\ a_{33} \end{bmatrix} = B_1^{-1} a_3 = \begin{bmatrix} 1 & -1/2 & 0 \\ 0 & 1/2 & 0 \\ 0 & 0 & 1 \end{bmatrix} \begin{bmatrix} 1 \\ -1 \\ 1 \end{bmatrix} = \begin{bmatrix} 3/2 \\ -1/2 \\ 1 \end{bmatrix}$$

$$E_1 = \begin{bmatrix} 1 & 0 & -3/2 \\ 0 & 1 & 1/2 \\ 0 & 0 & 1 \end{bmatrix}$$

$$B_2^{-1} = E_1 B_1^{-1} = \begin{bmatrix} 1 & 0 & -3/2 \\ 0 & 1 & 1/2 \\ 0 & 0 & 1 \end{bmatrix} \begin{bmatrix} 1 & -1/2 & 0 \\ 0 & 1 & 1/2 \\ 0 & 0 & 1 \end{bmatrix} =$$

$$\begin{bmatrix} 1 & -1/2 & -3/2 \\ 0 & 1/2 & 1/2 \\ 0 & 0 & 1 \end{bmatrix}$$

The Rest of problem proceeds as before.

Section 10.3

3.

x_1	3	0	0	3
x_2	2	1	0	1
x_3	2	0	1	0
x_4	0	2	0	3
x_5	0	1	1	2
x_6	0	0	2	1

$$\min z = x_1 + x_2 + x_3 + x_4 + x_5 + x_6$$
s.t. $3x_1 + 2x_2 + 2x_3 \geq 80$
$$x_2 + 2x_4 + x_5 \geq 50$$
$$x_3 + x_5 + 2x_6 \geq 100$$
$BV0 = \{x_1, x_4, x_6\}$

$$B_0 = \begin{bmatrix} 3 & 0 & 0 \\ 0 & 2 & 0 \\ 0 & 0 & 2 \end{bmatrix} \qquad B_0^{-1} = \begin{bmatrix} 1/3 & 0 & 0 \\ 0 & 1/2 & 0 \\ 0 & 0 & 1/2 \end{bmatrix}$$

$$c_{BV}B_0^{-1} = [1 \ 1 \ 1] \begin{bmatrix} 1/3 & 0 & 0 \\ 0 & 1/2 & 0 \\ 0 & 0 & 1/2 \end{bmatrix} = [1/3 \ 1/2 \ 1/2]$$

$$c_{BV}B_0^{-1} \begin{bmatrix} a_4 \\ a_6 \\ a_7 \end{bmatrix} - 1 = [1/3 \ 1/2 \ 1] \begin{bmatrix} a_4 \\ a_6 \\ a_7 \end{bmatrix} - 1 = 1/3 a_4 + 1/2 a_6 + a_7 - 1$$

max $(1/3)a_4 + (1/2)a_6 + (1/2)a_7 - 1$
s.t. $4a_4 + 6a_6 + 7a_7 \leq 15$
Optimal solution is z= $1/6$ $a_4 = 2$, $a_7 = 1$. Thus x_3 enters the basis.

$$x_3 \text{ column} = B_0^{-1} \begin{bmatrix} 2 \\ 0 \\ 1 \end{bmatrix} = \begin{bmatrix} 1/3 & 0 & 0 \\ 0 & 1/2 & 0 \\ 0 & 0 & 1/2 \end{bmatrix} \begin{bmatrix} 2 \\ 0 \\ 1 \end{bmatrix} = \begin{bmatrix} 2/3 \\ 0 \\ 1/2 \end{bmatrix}$$

$$\text{Right Hand Side} = B_0^{-1} b = \begin{bmatrix} 1/3 & 0 & 0 \\ 0 & 1/2 & 0 \\ 0 & 0 & 1/2 \end{bmatrix} \begin{bmatrix} 80 \\ 50 \\ 100 \end{bmatrix} = \begin{bmatrix} 80/3 \\ 25 \\ 50 \end{bmatrix}$$

Enter x_3 in Row 1 $BV(1) = \{x_3, x_4, x_6\}$

$$B_1^{-1} = \begin{bmatrix} 1/2 & 0 & 0 \\ 0 & 1/2 & 0 \\ -1/4 & 0 & 1/2 \end{bmatrix}$$

$$c_{BV}B_1^{-1} = [1 \ 1 \ 1] \begin{bmatrix} 1/2 & 0 & 0 \\ 0 & 1/2 & 0 \\ -1/4 & 0 & 1/2 \end{bmatrix} = [1/4 \ 1/2 \ 1/2]$$

$$[1/4 \ 1/2 \ 1] \begin{bmatrix} a_4 \\ a_6 \\ a_7 \end{bmatrix} -1 = (1/4)a_6 + (1/2)a_7 - 1$$

$\max z = (1/4)a_6 + (1/2)a_7 - 1$
  s.t. $4a_4 + 6a_6 + 7a_7 \le 15$
Optimal solution to this problem has $z = 0$, so we have solved the LP.

$$\begin{bmatrix} x_3 \\ x_4 \\ x_6 \end{bmatrix} = B_1^{-1}b = \begin{bmatrix} 1/2 & 0 & 0 \\ 0 & 1/2 & 0 \\ -1/4 & 0 & 1/2 \end{bmatrix} \begin{bmatrix} 80 \\ 50 \\ 100 \end{bmatrix} = \begin{bmatrix} 40 \\ 25 \\ 30 \end{bmatrix}$$
  $z = 95$

Section 10.4

1.
$\max z = 7x_1 + 5x_2 + 3x_3$
  s.t. $x_1 + 2x_2 + x_3 \le 10$ Centralized Constraint
    $x_1 \le 3$ Constraint Set 1
    $x_2 + x_3 \le 5$ Constraint Set 2
    $2x_2 + x_3 \le 8$
        $x_1, x_2, x_3 \le 0$

$[x_1] = u_1 [0] + u_2[3] = [3u_2]$

$$\begin{bmatrix} x_2 \\ x_3 \end{bmatrix} = \lambda_1 \begin{bmatrix} 0 \\ 0 \end{bmatrix} + \lambda_2 \begin{bmatrix} 4 \\ 0 \end{bmatrix} + \lambda_3 \begin{bmatrix} 3 \\ 2 \end{bmatrix} + \lambda_4 \begin{bmatrix} 0 \\ 5 \end{bmatrix} = \begin{bmatrix} 4\lambda_2 + 3\lambda_3 \\ 2\lambda_3 + 5\lambda_4 \end{bmatrix}$$

Objective function:
$7x_1 + 5x_2 + 3x_3 = 7(3u_2) + 5(4\lambda_2 + 3\lambda_3) + 3(2\lambda_3 + 5\lambda_4)$
        $= 21 u_2 + 20\lambda_2 + 21\lambda_3 + 15\lambda_4$

Centralized Constraint: $3u_2 + 8\lambda_2 + 8\lambda_3 + 5\lambda_4 \leq 10$
Restricted Master:

$\quad$ Max $z = 21u_2 + 20\lambda_2 + 21\lambda_3 + 15\lambda_4$
$\quad\quad$ s.t. $\quad 3u_2 + 8\lambda_2 + 8\lambda_3 + 5\lambda_4 + s_1 = 10$
$\quad\quad\quad\quad u_1 + u_2 = 1$
$\quad\quad\quad\quad\quad \lambda_1 + \lambda_2 + \lambda_3 + \lambda_4 = 1$
$\quad\quad\quad\quad u_i, \lambda_i \geq 0$
$\quad\quad$ BV(0) = $\{s_1, u_1, \lambda_1\}$

$$B_0^{-1} = \begin{bmatrix} 1 & 0 & 0 \\ 0 & 1 & 0 \\ 0 & 0 & 1 \end{bmatrix}$$

$$c_{BV} B_0^{-1} = [0\ \ 0\ \ 0] \begin{bmatrix} 1 & 0 & 0 \\ 0 & 1 & 0 \\ 0 & 0 & 1 \end{bmatrix} = [0\ \ 0\ \ 0]$$

Obj. Func. Coeff for $u_i = 7x_1$

$$\begin{array}{c} x_1 \\ \text{Column in Constraints for } u_i = 1 \\ 0 \end{array}$$

$$c_{BV} B_0^{-1} \begin{bmatrix} x_1 \\ 1 \\ 0 \end{bmatrix} \ -7x_1 = -7x_1$$

Min $z = -7x_1 \quad$ s.t. $x_1 \leq 3, x_1 \geq 0$
Opt: $z = -21, x_1 = 3 \Rightarrow u_2$ enters basis

Obj. Func. Coeff for $\lambda_i = 5x_2 + 3x_3$
$\quad\quad\quad\quad\quad\quad\quad 2x_2 + x_3$

Col. in Constraints for λ_i $\begin{bmatrix} 0 \\ 1 \end{bmatrix}$

$c_{BV} B_0^{-1} \begin{bmatrix} 2x_2 + x_3 \\ 0 \\ 1 \end{bmatrix} -(5x_2 + 3x_3) = -5x_2 - 3x_3$

Min z = - 5x_2 - 3x_3
  s.t. $x_2 + x_3 \leq 5$
         $2x_2 + x_3 \leq 8$ $x_3, x_2 \geq 0$
  Opt: z = -21, x_2 = 3, x_3 = 2 (λ_3)
Arbitrarily choose u_2 to enter. The u_2 column is

$B_0^{-1} \begin{bmatrix} 3 \\ 1 \\ 0 \end{bmatrix} = \begin{bmatrix} 3 \\ 1 \\ 0 \end{bmatrix}$ $B_0^{-1} b = \begin{bmatrix} 1 & 0 & 0 \\ 0 & 1 & 0 \\ 0 & 0 & 1 \end{bmatrix} \begin{bmatrix} 10 \\ 1 \\ 1 \end{bmatrix} = \begin{bmatrix} 10 \\ 1 \\ 1 \end{bmatrix}$

BV(1) = {s_1, u_2, λ_1}

$B_1^{-1} = \begin{bmatrix} 1 & -3 & 0 \\ 0 & 1 & 0 \\ 0 & 0 & 1 \end{bmatrix}$

$c_{BV} B_1^{-1} = [0\ 21\ 0] \begin{bmatrix} 1 & -3 & 0 \\ 0 & 1 & 0 \\ 0 & 0 & 1 \end{bmatrix} = [0\ 21\ 0]$

Subproblem 1: $c_{BV} \, B_1^{-1} \begin{bmatrix} x_1 \\ 1 \\ 0 \end{bmatrix} - 7x_1 = 21 - 7x_1$

Min $z = 21 - 7x_1$ s.t. $x_1 \le 3$, $x_1 \ge 0$
Opt. $z = 0$, $x_1 = 3$

Subproblem 2: $c_{BV} \, B_1^{-1} \begin{bmatrix} 2x_2 & + & x_3 \\ & 0 & \\ & 1 & \end{bmatrix} - 5x_2 - 3x_3 = -5x_2 - 3x_3$

Same LP as last phase => enter λ_3

$$B_1^{-1} \begin{bmatrix} 2(3) & + & 2 \\ & 0 & \\ & 1 & \end{bmatrix} = \begin{bmatrix} 1 & -3 & 0 \\ 0 & 1 & 0 \\ 0 & 0 & 1 \end{bmatrix} \begin{bmatrix} 8 \\ 0 \\ 1 \end{bmatrix} = \begin{bmatrix} 8 \\ 0 \\ 1 \end{bmatrix}$$

$$B_1^{-1} b = \begin{bmatrix} 1 & -3 & 0 \\ 0 & 1 & 0 \\ 0 & 0 & 1 \end{bmatrix} \begin{bmatrix} 10 \\ 1 \\ 1 \end{bmatrix} = \begin{bmatrix} 7 \\ 1 \\ 1 \end{bmatrix}$$

$$BV(2) = \{\lambda_3, u_2, \lambda_1\} \qquad B_2^{-1} = \begin{bmatrix} 1/8 & -3/8 & 0 \\ 0 & 1 & 0 \\ -1/8 & 3/8 & 1 \end{bmatrix}$$

$$c_{BV} \, B_2^{-1} = [21 \ \ 21 \ \ 0] \begin{bmatrix} 1/8 & -3/8 & 0 \\ 0 & 1 & 0 \\ -1/8 & 3/8 & 1 \end{bmatrix} = [21/8 \ \ 105/8 \ \ 0]$$

Sub-problem 2: $c_{BV} \, B_2^{-1} \begin{bmatrix} 2x_2 \ + \ x_3 \\ 0 \\ 1 \end{bmatrix} - 5x_2 - 3x_3 =$

$21/8 \, (2x_2 + x_3) - 5x_2 - 3x_3 = 1/4 \ x_2 - 3/8 \ x_3$
  $\min z = 1/4 x_2 - 3/8 x_3$
   s.t. $\quad x_2 + x_3 \leq 5$
   $\quad\quad 2x_2 + x_3 \leq 8$
   $\quad\quad\quad\quad\quad\quad x_2, x_3 \geq 0$
  Opt: $z = -15/8$, $x_2 = 0$, $x_3 = 5$
   $\Rightarrow$ Enter λ_4

$$B_2^{-1} \begin{bmatrix} 2(0) \ + \ 5 \\ 0 \\ 1 \end{bmatrix} = \begin{bmatrix} 1/8 & -3/8 & 0 \\ 0 & 1 & 0 \\ -1/8 & 3/8 & 1 \end{bmatrix} \begin{bmatrix} 5 \\ 0 \\ 1 \end{bmatrix} = \begin{bmatrix} 5/8 \\ 0 \\ 3/8 \end{bmatrix}$$

$$B_2^{-1} b = \begin{bmatrix} 1/8 & -3/8 & 0 \\ 0 & 1 & 0 \\ -1/8 & 3/8 & 1 \end{bmatrix} \begin{bmatrix} 10 \\ 1 \\ 1 \end{bmatrix} = \begin{bmatrix} 7/8 \\ 1 \\ 1/8 \end{bmatrix}$$

$$BV(3) = \{\lambda_3, u_2, \lambda_4\} \; ; B_3^{-1} = \begin{bmatrix} 1/3 & -1 & -5/3 \\ 0 & 1 & 0 \\ -1/3 & 1 & 8/3 \end{bmatrix}$$

$$c_{BV} B_3^{-1} = [21 \ 21 \ 15] \begin{bmatrix} 1/3 & -1 & -5/3 \\ 0 & 1 & 0 \\ -1/3 & 1 & 8/3 \end{bmatrix} = [2 \ 15 \ 5]$$

Subproblem 2: $c_{BV} B_3^{-1} \begin{bmatrix} 2x_2 + x_3 \\ 0 \\ 1 \end{bmatrix} - 5x_2 - 3x_3 = 5 - x_2 - x_3$

Min $z = 5 - x_2 - x_3$ s.t. $x_2 + x_3 \le 5$
$\qquad\qquad\qquad\qquad 2x_2 + x_3 \le 8 \qquad x_2, x_3 \ge 0$

Opt: $z = 0 \quad x_2 = 3, x_3 = 2 \qquad$ or $x_2 = 0, x_3 = 5$

Optimal Solution has been found.

Section 10.5

1.
$\max z = 4x_1 + 3x_2 + 5x_3$
$\quad$ s.t. $\ 2x_1 + 2x_2 + x_3 + x_4 \le 9$
$\qquad\quad 4x_1 - x_2 - x_3 + x_5 \le 6$
$\qquad\quad 2x_2 + x_3 \le 5$
$\quad x_1 \le 2, x_2 \le 3, x_3 \le 4, x_4 \le 5, x_5 \le 7$
$\quad x_1, x_2, x_3, x_4, x_5 \ge 0$

Initial Tableau:
$\quad z - 4x_1 - 3x_2 - 5x_3 = 0 \qquad\qquad\qquad z = 0$
$\qquad 2x_1 + 2x_2 + x_3 + x_4 + s_1 = 9 \qquad\quad s_1 = 9$
$\qquad 4x_1 - x_2 - x_3 + x_5 + s_2 = 6 \qquad\quad s_2 = 6$
$\qquad 2x_2 + x_3 + s_3 = 5 \qquad\qquad\qquad\quad s_3 = 5$
For x_3: $BN_1 = 4 \qquad s_1 = 9 - x_3$ $(s_1 \ge 0$ iff $x_3 \le 9)$
$\qquad\qquad\qquad\qquad s_2 = 6 + x_3$ $(s_2 \ge 0$ iff $x_3 \ge -6)$
$\qquad\qquad\qquad\qquad s_3 = 5 - x_3$ $(s_3 \ge 0$ iff $x_3 \le 5)$ => $BN_2 = 5$
$\quad$ replace x_3 with $4 - x_3'$
$\quad z - 4x_1 - 3x_2 + 5x_3' = 20 \qquad\qquad z = 20$
$\qquad 2x_1 + 2x_2 - x_3' + x_4 + s_1 = 5 \qquad\quad s_1 = 5$

76

$$4x_1 - x_2 + x_3' + x_5 + s_2 = 10 \qquad s_2 = 10$$
$$2x_2 - x_3' + s_3 = 1 \qquad s_3 = 1$$

For x_1: $BN_1 = 2$ $\qquad s_1 = 5 - 2x_1$ ($s_1 \geq 0$ iff $x_1 \leq 5/2$)
$\qquad\qquad\qquad s_2 = 10 - 4x_1$ ($s_2 \geq 0$ iff $x_1 \leq 5/2$) $\Rightarrow BN_2 = 5/2$
replace x_1 by $2 - x_1'$

$$z + 4x_1' - 3x_2 + 5x_3' = 28 \qquad z = 28$$
$$-2x_1' + 2x_2 - x_3' + x_4 + s_1 = 1 \qquad s_1 = 1$$
$$-4x_1' - x_2 + x_3' + x_5 + s_2 = 2 \qquad s_2 = 2$$
$$2x_2 - x_3' + s_3 = 1 \qquad s_3 = 1$$

For x_2: $BN_1 = 3$ $\qquad s_1 = 1 - 2x_2$ ($s_1 \geq 0$ iff $x_2 \leq 1/2$)
$\qquad\qquad\qquad s_2 = 2 + x_2$ ($s_2 \geq 0$ iff $x_2 \geq -2$)
$\qquad\qquad\qquad s_3 = 1 - 2x_2$ ($s_1 \geq 0$ iff $x_2 \leq 1/2$) $\Rightarrow BN_2 = 1/2$
Enter x_2 in Row 3
$$z + 4x_1' + 7/2x_3' + 3/2s_3 = 29\ 1/2 \qquad\qquad z = 210.5$$
$$-2x_1' + x_4 + s_1 - s_3 = 0 \qquad\quad s_1 = 1$$
$$-4x_1' + 1/2x_3' + x_5 + s_2 + 1/2s_3 = 5/2 \quad s_2 = 5/2$$
$$x_2 - 1/2x_3' + 1/2s_3 = 1/2 \qquad\quad x_2 = 1/2$$

Optimal Solution:
$z = 29.5$
$s_1 = 0$, $s_2 = 5/2$, $x_2 = 1/2$, $s_3 = 0$, $x_1' = 0$, $x_3' = 0$, $x_4 = 0$, $x_5 = 0$
$x_1 = 2 - x_1' = 2 - 0 = 2$
$x_3 = 4 - x_3' = 4 - 0 = 4$

Section 10.6

1. Choose $\epsilon = .1$. $x^0 = [1/3\ 1/3\ 1/3]^T$ and $k = 0$. Since $z = 2/3 > .1$, we proceed to Step 3.

$$A = [1\ 0\ -1] \qquad \text{Diag } x^0 = \begin{bmatrix} 1/3 & 0 & 0 \\ 0 & 1/3 & 0 \\ 0 & 0 & 1/3 \end{bmatrix}$$

$$A[\text{Diag}(x^0)] = [1/3\ 0\ -1/3]$$

$$P = \begin{bmatrix} 1/3 & 0 & -1/3 \\ 1 & 1 & 1 \end{bmatrix}$$

$$PP^T = \begin{bmatrix} 2/9 & 0 \\ 0 & 3 \end{bmatrix}$$

$$(PP^T)^{-1} = \begin{bmatrix} 9/2 & 0 \\ 0 & 1/3 \end{bmatrix}$$

$$(I - P^T(PP^T)^{-1}P) = \begin{bmatrix} 1/6 & -1/3 & 1/6 \\ -1/3 & 2/3 & -1/3 \\ 1/6 & -1/3 & 1/6 \end{bmatrix}$$

$c = [1\ 2\ -1]$

$[\text{Diag } x^0]c^T = [1/3\ 2/3\ -1/3]^T$

$(I - P^T (PP^T)^{-1}P)[\text{Diag } x^0]c^T = [-2/9\ 4/9\ -2/9]^T$

Using $\theta = .25$ we obtain $y^1 = [3/8\ 1/4\ 3/8]^T$ and $x^1 = [3/8\ 1/4\ 3/8]^T$. Since $z = .5 > .1$, another iteration would be necessary.

Chapter 11 Solutions

Section 11.1

1. $\lim\limits_{h \to 0} \dfrac{3h + h^2}{h} = \lim\limits_{h \to 0} (3 + h) = 3$

3a. $x(-e^{-x}) + e^{-x}$

4. $\partial f/\partial x_1 = 2x_1 \exp(x_2)$ $\partial f/\partial x_2 = x_1^2 \exp(x_2)$ $\partial f^2/\partial x_1 \partial x_2 =$ $\partial f^2/\partial x_2 \partial x_1 = 2x_1 \exp(x_2) = \partial^2 f/\partial^2 x_1 = 2\exp(x_2), \partial^2 f/\partial^2 x_2 = x_1^2 \exp(x_2).$

Section 11.2

1a. Let S = soap opera ads and F = football ads. Then we wish to

min $z = 50S + 100F$

st $\quad 5S^{1/2} + 17F^{1/2} \geq 40$ (men)

$\quad\quad 20S^{1/2} + 7F^{1/2} \geq 60$ (women)

$\quad\quad S \geq 0,\ F \geq 0$

1b. Since doubling S does not double the contribution of S to each constraint, we are violating the proportionality assumption. Additivity is not violated.

1c. This accounts for the fact that an extra soap opera ad yields a benefit which is a decreasing function of the number of football ads. This accounts for the fact that we may not want to double count people who see both types of ads.

4. Let S = number of soap opera ads and F = number of football ads. See LINGO printout for solution.

LINGO Printout for Problem 4 Section 11.2

```
MODEL:
MIN=50*S+100*F;
5*S^.5+17*F^.5>40;
20*S^.5+7*F^.5>60;
S>0;
F>0;
END
```

Local optimal solution found at step: 21
Objective value: 563.0744

Variable	Value
S	5.886590
F	2.687450

Reduced Cost

0.0000000

0.0000000

Row	Slack or Surplus
1	563.0744
2	0.0000000 -
3	0.0000000 -
4	5.886590
5	2.687450

Dual Price

1.000000

15.93120

8.148348

0.0000000

0.0000000

Section 11.3

2. $f''(x)>0$ for $x>0$ and $f''(x)<0$ for $x<0$, so $f(x)$ is neither convex nor concave.

5. $f''(x) = -x^{-2}<0$, so $f(x)$ is a concave function on S.

10. $H = \begin{bmatrix} 2a & b \\ b & 2c \end{bmatrix}$

The function will be convex if $2c\geq0$, $a\geq0$, and $4ac\geq b^2$. These conditions ensure that all principal minors will have nonnegative determinants. The function will be concave if $a\leq0$, $2c\leq0$ and

80

4ac - $b^2 \geq 0$. These conditions ensure that both principal minors are nonpositive and the second principal minor is nonnegative.

Section 11.4

1. Let $f(x)$ = profit if $x is spent on advertising. Then
$f(0) = 0$ and for $x>0$, $f(x) = 300x^{1/2} - 100x^{1/2} - 5000 - x$.
Since $f(x)$ has no derivative at $x = 0$, maximum profit occurs either for $x = 0$ or a point where $f'(x) = 0$.
Now for $x>0$ $f'(x) = 100x^{-1/2} - 1 = 0$ for $x = 10,000$. Also
$f''(x) = -50x^{-3/2}<0$ for $x>0$. Thus $x = 10,000$ is a local maximum(and a maximum over all $x>0$). We now compare $f(0)$ and $f(10,000)$ to determine what the company should do. $f(0) = 0$ and $f(10,000) = $5,000$, so company should spend $10,000 on advertising.
  If fixed cost is $20,000, $f'(x) = 0$ still holds for $x = 10,000$. Comparing $f(0) = 0$ and $f(10,000) = -10,000$, we now find that $x = 0$ is optimal.

6. $f'(x) = 3x^2 - 6x + 2$, $f''(x) = 6x - 6$.
The quadratic formula yields $f'(.42) = f'(1.58) = 0$. Since $f''(1.58)>0$ and $f''(.42)<0$ we know that $x = 1.58$ is a local minimum. Thus extremum candidates are -2, 4, and 1.58. $f'(-2)>0$, so $x = -2$ is a local minimum. $f'(4)>0$, so $x = 4$ is a local maximum. Since $f(-2) = -25$ and $f(1.58) = -1.39$ we find that the NLP is solved by $x = -4$.

Section 11.5

1. $a = -3$ $b = 5$, so $b - a = 8$. $x_1 = 5 - .618(8) = .056$
$x_2 = -3 + .618(8) = 1.944$. $f(x_1) = .115$, $f(x_2) = 7.67$
$f(x_2)>f(x_1)$ so interval of uncertainty is now $(.056,5]$. Then
$x_3 = x_2$, $x_4 = .056 + .618(4.944) = 3.11$.
$f(x_4) = 15.89>f(x_3) = f(x_2) = 7.67$. Thus new interval of uncertainty is $(1.944,5]$. Now $x_5 = x_4 = 3.11$ and
$x_6 = 1.944 + .618(3.056) = 3.83$. $f(x_5) = 15.89$, $f(x_6) = 22.33$.
Since $f(x_6)>f(x_5)$ new interval of uncertainty is $(3.11,5]$.
$x_7 = x_6 = 3.83$ and $x_8 = 3.11 + .618(1.89) = 4.28$.
$f(x_8) = 26.88$ and $f(x_8)>f(x_7)$. Thus new interval of uncertainty is $(3.83,5]$. $x_9 = x_8 = 4.28$ and $x_{10} = 3.83 + .618(1.17) = 4.55$.

$f(x_{10}) = 29.8 > f(x_9)$ so new interval of uncertainty is $(4.28,5]$. This interval has length $.72 < .8$. Thus we know that maximum occurs for some value of x on interval $(4.28,5]$. (maximum actually occurs for $x = 5$).

Section 11.6

3. Let $f(q_1, q_2) = $ profit when company 1 sells q_1 units and company 2 sells q_2 units. Then

$$f(q_1, q_2) = (q_1 + q_2)(200 - q_1 - q_2) - q_1 - .5q_2^2$$
$$= 199q_1 + 200q_2 - q_1^2 - 2q_1q_2 - 1.5q_2^2$$
$\partial f/\partial q_1 = 199 - 2q_1 - 2q_2$ and $\partial f/\partial q_2 = 200 - 3q_2 - 2q_1$.
$\partial f/\partial q_1 = \partial f/\partial q_2 = 0$ for $q_1 = 98.5$ and $q_2 = 1$. Since

$$H = \begin{bmatrix} -2 & -2 \\ -2 & -3 \end{bmatrix} \quad \det H_1 = -2 < 0, \det H_2 = 2 > 0.$$

Thus the above values of q_1 and q_2 are a local max. Since $f(q_1, q_2)$ is a concave function, we know that $(98.5, 1)$ maximizes profit over all values of q_1 and q_2.

Section 11.7

2. $\nabla f(x_1, x_2) = [3 - 2x_1 \ -2x_2]$
  $\nabla(2.5, 1.5) = [-2, -3]$. To find a new point solve
max $-(0.5 - 2t)^2 - (2.5 - 2t) - (1.5 - 3t)^2 = f(t)$
$t \geq 0$
    $f'(t) = 4(0.5 - 2t) + 2 + 6(1.5 - 3t) = 0$ if
      $13 - 26t = 0$ or $t = .50$
New Point $= (2.5, 1.5) + .5(-2, -3) = (1.5,0)$
Since $\nabla(1.5, 0) = [0 \ 0]$ we conclude the algorithm.

Section 11.8

2. We wish to maximize $L^{2/3}K^{1/3}$ subject to $2L + K = 10$. It is easier to maximize $\ln L^{2/3}K^{1/3} = (2/3)\ln L + (1/3)\ln K$

82

(this is a concave function so we know that Lagrange multipliers will yield a maximum). Forming the Lagrangian LAG we find that

LAG = (2/3)ln L + (1/3)ln K + λ(10 - 2L - K)
(1) ∂LAG/∂L = 2/3L - 2λ = 0 , (2) ∂LAG/∂K = 1/3K - λ = 0
(3) ∂LAG/∂λ = 10 - 2L - K = 0 From (1) L = 1/3λ.
From (2) K = 1/3λ and from (3) 2(1/3λ) + (1/3λ) = 10 or
λ = 1/10. Then L = K = 10/3.

3. min 2L + K
   st $L^{2/3}K^{1/3}$ = 6 (or (2/3)ln L + (1/3)ln K = ln 6)
The Lagrangian (LAG) is given by
LAG = 2L + K +λ(ln 6 - (2/3)ln L - (1/3)ln K)

(1) ∂LAG/∂L = 2 - 2λ/3L = 0, (2) ∂LAG/∂K = 1 - λ/3K = 0
(3) ∂LAG/∂λ = (2/3)ln L + (1/3)ln K - ln 6 = 0
(1) and (2) yield L = K = λ/3. Then (3) yields λ = 18, and (1) and (2) yield L = K = 6 .

Section 11.9

1. max p_1(60 - .5p_1) + p_2(40 - p_2) - 10c
   st 60 - .5p_1 - c≤0
      40 - p_2 - c≤0
         All variables ≥0
(c = generating capacity)
Ignoring the non-negativity restrictions the K-T conditions consist of the original constraints and
(1) 60 - p_1 + .5$λ_1$ = 0 (p_1 constraint)
(2) 40 - 2p_2 + $λ_2$ = 0 (p_2 constraint)
(3) -10 + $λ_1$ + $λ_2$ = 0 (c constraint)
(4) $λ_1$(.5p_1 + c - 60) = 0
(5) $λ_2$(p_2 + c - 40) = 0 $λ_1$, $λ_2$≥0
K-T conditions have a solution where $λ_1$>0 and $λ_2$ = 0. Then (3) yields $λ_1$ = 10. Now (1) yields p_1 = 65 and (2) yields p_2 = 20. Finally, (4) yields c = 27.5. Since objective function is concave and constraints are linear, we have found the optimal solution.

6. Replace the constraint $-x_1 + x_2 = 1$ by $-x_1 + x_2 \leq 1$ and $x_1 - x_2 \leq -1$. Then (29)-(33)' yield the following K-T conditions.

(1) $2(x_1 - 1) - \lambda_1 + \lambda_2 + \lambda_3 \geq 0$ (x_1 constraint)

(2) $2(x_2 - 2) + \lambda_1 - \lambda_2 + \lambda_3 \geq 0$ (x_2 constraint)

(3) $\lambda_1(1 + x_1 - x_2) = 0$

(4) $\lambda_2(-1 - x_1 + x_2) = 0$

(5) $\lambda_3(2 - x_1 - x_2) = 0$

(6) $x_1(2(x_1 - 1) - \lambda_1 + \lambda_2 + \lambda_3) = 0$

(7) $x_2(2(x_2 - 2) + \lambda_1 - \lambda_2 + \lambda_3) = 0$

Try $\lambda_3 > 0$. Then (5) yields $x_1 + x_2 = 2$. We know, however, that $-x_1 + x_2 = 1$. Solving simultaneously yields $x_1 = 1/2$, $x_2 = 3/2$. If we now try $\lambda_1 = \lambda_2 = 0$ (6) and (7) yield $\lambda_3 = 1$. All the K-T conditions are now satisfied. Since the objective function is convex and the constraints are linear, we have found an optimal solution to the NLP.

Section 11.10

1. Let x_i = amount invested in stock. Then variance of portfolio
= var $(x_1 S_1 + x_2 S_2 + x_3 S_3)$
= x_1^2var S_1 + x_2^2var S_2 + x_3^2varS$_3$ + $2x_1 x_2$cov(S_1, S_2)
+ $2x_1 x_3$cov(S_1, S_3) + $2x_2 x_3$cov(S_2, S_3) = $.09x_1^2$ + $.04x_2^2$
+ $.01x_3^2$ + $.012x_1 x_2$ - $.008x_1 x_3$ + $.010x_2 x_3$

Thus the appropriate NLP is

min $z = .09x_1^2 + .04x_2^2 + .01x_3^2 + .012x_1 x_2 - .008x_1 x_3 + .010x_2 x_3$
st $\quad \dfrac{.15x_1 + .21x_2 + .09x_3}{x_1 + x_2 + x_3} \geq .15$

or $.06x_2 - .06x_3 \geq 0$ or $x_2 - x_3 \geq 0$
$\quad x_1 + x_2 + x_3 = 100$
$\quad x_1, x_2, x_3 \geq 0$

max $p_1(4000 - 10p_1 + p_2) + p_2(2000 - 9p_2 + .8p_1)$
$\quad$ s.t $2(4000 - 10p_1 + p_2) + 3(2000 - 9p_2 + .8p_1) \leq 5000$ (labor)
$\quad\quad 3(4000 - 10p_1 + p_2) + 2000 - 9p_2 + .8p_1 \leq 4500$ (chips)
$\quad\quad p_1, p_2 \geq 0$

84

LINGO yields an optimal solution of z = -999,535, P1 = $292.81, P2 = $158.33, M1 = 0, M2 = $53.81. Thus adding an additional labor hour will not increase revenue, so Fruit would be willing to pay $0 for an additional hour of labor. If one more chip were available, then revenue is increased by approximately $53.81, so Fruit would pay up to (slightly less than) $53.81 for another chip. Total revenue of $999,535 will be earned.

Section 11.11

1. We know that in the optimal solution x_1 is between 0 and 3 and x_2 is between 0 and 2. Thus we may choose the following grid points: For x_1: 0, 1, 2, 3 For x_2 0, 2/3, 4/3, 2 Then the approximating problem is

$$\min z = \delta_{12} + 4\delta_{13} + 9\delta_{14} + 4\delta_{22}/9 + 16\delta_{23}/9 + 4\delta_{24}$$

st $\delta_{12} + 4\delta_{13} + 9\delta_{14} + 2(4\delta_{22}/9 + 16\delta_{23}/9 + 4\delta_{24}) \leq 4$

$\delta_{12} + 4\delta_{13} + 9\delta_{14} + 4\delta_{22}/9 + 16\delta_{23}/9 + 4\delta_{24} \leq 6$

$\delta_{11} + \delta_{12} + \delta_{13} + \delta_{14} = 1$, $\delta_{21} + \delta_{22} + \delta_{23} + \delta_{24} = 1$

Adjacency assumption plus all variables ≥ 0

Our approximating problem yields the following values for x_1 and x_2:
$x_1 = \delta_{12} + 2\delta_{13} + 3\delta_{14}$, $x_2 = 2\delta_{22}/3 + 4\delta_{23}/3 + 2\delta_{33}$

Section 11.12

1. $x^0 = [1/2 \ 1/2]^T$ $\nabla f(x,y) = [4 - 4x - 2y, 6 - 2x - 4y]$
$\nabla f(.5,.5) = [1 \ 3]$ Find d^0 by solving
$$\max z = d1 + 3d2$$
st $d1 + 2d2 \leq 2$
$$d1, d2 \geq 0$$
Optimal solution is $d^0 = [0 \ 1]T$ Choose $x^1 = [.5 \ .5]^T + t_0[-.5 \ .5]^T$
$= [.5 - .5t_0 \ .5 + .5t_0]$ where t_0 solves
$\max$ $f(.5 - .5t, .5 + .5t) = 3.5 + t - t^2/2 = g(t)$.
$0 \leq t \leq 1$
Then $g'(t) = 1 - t = 0$ for $t = 1$. Since $g''(t) < 0$, $t_0 = 1$ and
$x^1 = [0 \ 1]^T$. Here $z = f(0, 1) = 4$. $\nabla f(x^1) = [2 \ 2]$. We find d^1 by solving
$$\max z = 2d1 + 2d2$$
st $d1 + 2d2 \leq 2$
$$d1, d2 \geq 0$$

85

Optimal solution is $d^1 = [2 \ 0]^T$. Now $x^2 = [0 \ 1]^T + t_1[2 \ -1]^T$
$= [2t_1 \ 1 - t_1]$ where t_1 solves
max $f(2t, 1 - t) = 4 + 2t - 6t^2 = h(t)$. Then $h'(t) = 2 - 12t = 0$
$0 \le t \le 1$
for $t = 1/6$. Since $h''(t) < 0$, $t_1 = 1/6$ and $x^2 = [1/3 \ 5/6]$. At this point $z = 4.17$.

Section 11.13

We first solve LP to maximize profit

$$\max z = 500x1 + 1100x2 - 10(OT)$$
$$st \quad 6x1 + 12x2 \le 200$$
$$8x1 + 20x2 \le 300$$
$$11x1 + 24x2 \le 300 + OT$$
$$x1, x2, OT \ge 0$$

Here x_i = Units of product I produced and OT = hours of overtime used. Optimal solution has Profit = \$16,666.67 and OT = 83.33. We now add constraint OT$\le$k, and set k = 80, 75, 70, 65, 60, 50, 40, ... 0 and obtain the given tradeoff curve. For instance, when k = 10 optimal z-value is \$14,108.33, so (10, 14,108.33) is on tradeoff curve.

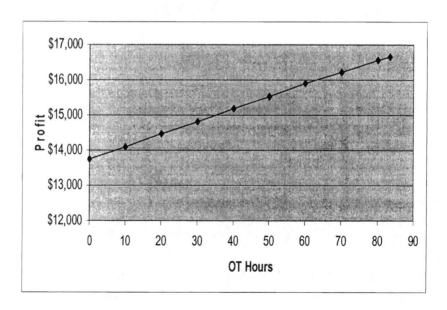

Chapter 12 Solutions

SECTION 12.1

1. $\int_{0}^{5} e^{2t}dt = [e^{2t}/2]_{0}^{5} = \dfrac{e^{10} - 1}{2}$

SECTION 12.2

1. $F'(y) = 2y(2y + y^2) - (1)(3y) + \int_{y}^{y^2} 2dx$

$\qquad = 2y^3 + 4y^2 - 3y + 2(y^2 - y)$

SECTION 12.3

1a. Sample space consists of 36 equally likely points $(1,1)$, $(1,2)$, ...$(1,6)$... $(6,1)$,... $(6,6)$, where point (i,j) means i dots show on first toss and j dots show on second toss.

Each point in the sample space has a probability 1/36 and six points yield a seven, so P(seven) = 6/36 = 1/6. Two points in sample space yield an eleven so P(eleven) = 2/36 = 1/18. Since a seven and an eleven are mutually exclusive P(seven or eleven) = 6/36 + 2/36 = 8/36

1b. Let $\overline{E}$ = event that a two or twelve is rolled. We seek P(E).
Now P(E) = P(two) + P(twelve) = 1/36 + 1/36 = 2/36 so

P($\overline{E}$) = 1 - 2/36 = 34/36 = 17/18.

$$P(E_1 \cap E_2)$$

1c. $P(E_1) = 1/6, P(E_2) = 5/36, \ P(E_2|E_1) = \ \dfrac{}{P(E_1)}$

$= (1/36)/(1/6) = 1/6 \neq P(E_2)$. Thus events E_1 and E_2 are not independent.

1d. $P(E_1) = 1/6, \ P(E_2) = 1/6, \ P(E_1 \cap E_2) = 1/36.$

$$P(E_2|E_1) = \frac{P(E_1 \cap E_2)}{P(E_2)} = \frac{1/36}{1/6} = 1/6.$$

Thus E_1 and E_2 are independent events.

1e. E_1 = event that total is five, E_2 = event that first die shows two dots. We seek $P(E_2|E_1)$.
$P(E_1 \cap E_2) = 1/36, \ P(E_1) = 4/36$
yields that
$$P(E_2|E_1) = P(E_1 \cap E_2)/P(E_1) = 1/4.$$

1f. E_1 = event that first die is a five, E_2 = event that total is even. Then
$$P(E_2|E_1) = \frac{P(E_1 \cap E_2)}{P(E_1)} = \frac{3/36}{6/36} = 1/2$$

SECTION 12.4

1. Let S_i = event that drawer i is chosen, O_1 = event that a silver coin is drawn, and O_2 = event that a gold coin is drawn.
We seek $P(S_3|O_1)$. We begin by determining the joint probabilities involving O_1:

$P(S_1 \cap O_1) = P(O_1|S_1)P(S_1) = (0)(1/3) = 0$

$P(S_2 \cap O_1) = P(O_1|S_2)P(S_2) = (1/2)(1/3) = 1/6$

$P(S_3 \cap O_1) = P(O_1|S_3)P(S_3) = 1(1/3) = 1/3$

Thus $P(O_1) = 0 + 1/6 + 1/3 = 1/2$ and

$$P(S_3|O_1) = \frac{P(O_1 \cap S_3)}{P(O_1)} = \frac{1/3}{1/2} = 2/3$$

SECTION 12.5

1a. Let S = number of items sold. Then $P(S = 90) = 1/3$ and $P(S = 100) = 2/3$. Thus $E(S) = 1/3(90) + 2/3(100) = 290/3$ and
var $S = (1/3)(90)^2 + (2/3)(100)^2 - (290/3)^2 = 200/9$.

1b. Let U = unfilled demand. Then $P(U = 0) = 2/3$ and $P(U = 10) = 1/3$. Thus $E(U) = 2/3(0) + 1/3(10) = 10/3$ and
var $U = 2/3(0)^2 + 1/3(10)^2 - (10/3)^2 = 200/9$.

6. $E(X) = 0$ $E(Y) = 2/3$.
cov $(X, Y) = 1/3\{(-1)(1/3) + (0)(-2/3) + (1)(1/3)\} = 0$.
Since $P(Y = 0 \,|X = 0) = 1$ and $P(Y = 0) = 1/3$, X and Y are not independent random variables.

SECTION 12.6

1. Let D = demand for milk and g = gallons of milk in stock. Then $P(D \geq g) = .05$ or

$$P\left(\frac{D - 1000}{10} \geq \frac{g - 1000}{10} \right) = .05$$

or $P\left(Z \geq \dfrac{g - 1000}{10}\right) = .05$. Using $F(1.65) = .95$

Thus

$$\frac{g - 1000}{10} = 1.65 \text{ or } g = 1016.5 \text{ gallons}$$

2. Let X_i = life of bulb i and T = total life of three bulbs.
Then $T = X_1 + X_2 + X_3$ and T is N(1500, 1200). Then

$$P(T \geq 1460) = P\left(\frac{T - 1500}{\sqrt{1200}} \geq \frac{1460 - 1500}{\sqrt{1200}}\right)$$

$$= P(Z \geq -1.15) = P(Z \leq 1.15) = .8749$$

Section 12.7

1. $P_X^T(z) = \sum_{n=0}^{n=\infty} e^{-\mu} \frac{\mu^n}{n!} z^n = e^{-\mu} \sum_{n=0}^{n=\infty} \frac{(\mu z)^n}{n!} = e^{\mu z - \mu}$.

1st Derivative of z-transform is $\mu e^{\mu z - \mu}$.

Thus setting z = 1 yields mean = μ.

2nd derivative of z-transform is $\mu^2 e^{\mu(z-1)}$.

Setting z = 1 we find $E(X^2) = \mu^2 + \mu$. Therefore variance of X is given by μ.

Chapter 13 Solutions

SECTION 13.1

1. For Maximax decision look at

Pizza King Chooses	Min Reward
Small	2,000*
Medium	1,000
Large	0

Thus maximax decision is to mount a small advertising campaign.
  For Maximax decision look at

Pizza King Chooses	Max Reward
Small	6,000
Medium	6,000
Large	9,000*

Thus maximax decision is to mount a large advertising campaign.
To find Minimax Regret decision compute regret matrix

		Noble Greek		
	Small	Medium	Large	Maximum Regret
Pizza King	Small	3,000	1,000	0

3,000

                 Medium   4,000   0   1,000
4,000

                 Large   0   0   2,000
2,000 *

Thus Minimax Regret decision is to choose a large advertising campaign.

SECTION 13.2

1a. $u'(x) = 1/x$, $u''(x) = -1/x^2 < 0$ so I am risk averse.

1b. $E(U \text{ for } L_1) = \ln(19,000) = 9.8521943$, $E(U \text{ for } L_2) = .1 (\ln 10,000) + .9(\ln 20,000) = 9.8341728$. Thus I prefer L_1.

EV for $L_2 = .1(10,000) + .9(20,000) = \$19,000$. Let $x = CE(L_2)$ (measured in terms of asset position). Then $\ln x = 9.8341728$ and $x = \$18,661$. Thus RP for $L_2 = \$19,000 - \$18,661 = \$339$.

2. $u'(x) = 2x$, $u''(x) = 2 > 0$, so I am risk seeking.
$E(U \text{ for } L_1) = (19,000)^2 = = 361,000,000$
$E(U \text{ for } L_2) = .1(10,000)^2 + .9(20,000)^2 = 370,000,000$.
Thus I now prefer L_2.
Let $x = CE(L_2)$. Then $x^2 = 370,000,000$ or $x = \$19,235$.
Since EV for $L_2 = \$19,000$ we find that
RP for $L_2 = 19,000 - 19,235 = -\$235$.

3. Since $u''(x) = 0$ I am now risk neutral.
$E(U \text{ for } L_1) = 2(19,000) + 1 = 38,001$
$E(U \text{ for } L_2) = .1(20,001) + .9(40,001) = 38,001$
   Thus I am indifferent between L_1 and L_2. (This is because a risk neutral decision maker chooses between lotteries on basis of expected value and L_1 and L_2 have same expected value. Let

$x = CE(L_2)$. Then $2x + 1 = 38,001$ or $x = \$19,000$. Thus

RP for $L_2 = 19,000 - 19,000 = \$0$.

SECTION 13.3

1.   We need to show that prospect theory (unlike Expected utility theory) allows the possibility that L_1 is preferred to L_2 and L_4 is preferred to L_3. Without loss of generality we set $u(5) = 1$ and $u(0) = 0$. Suppose $u(1) = .9$ and $\pi(.10) = .098$ $\pi(.89) = .89$ and $\pi(.11) = .099$. Then L_1 is preferred to L_2 because $u(1) > \pi(.10)*u(5) + \pi(.89)*u(1)$. Also L_4 is preferred to L_3 because $\pi(.10) > \pi(.11)u(1)$. Thus the preferences exhibited by most decision makers are consistent with prospect theory.

SECTION 13.4

1. Hire geologist. If favorable report we drill; if unfavorable report do not drill. Expected net profits = \$180,000. EVWSI = \$190,000 EVWOI = \$170,000. EVSI = \$190,000 - \$170,000 = \$20,000. Since EVSI> cost of geologist we should hire geologist.

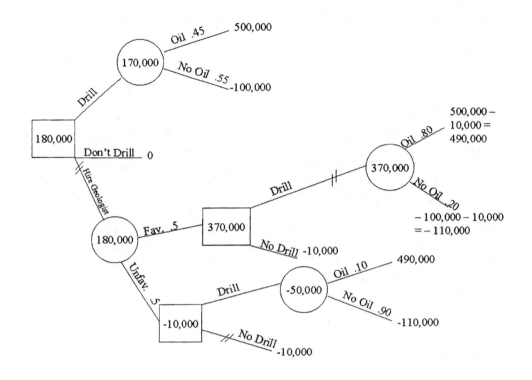

2. We maximize revenues less costs = -costs

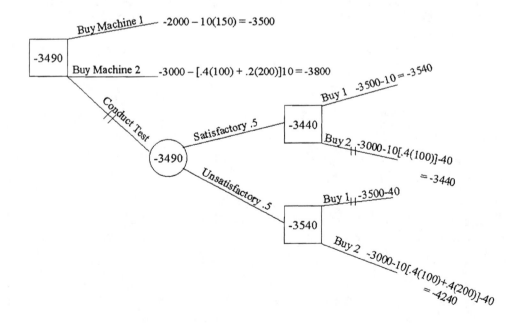

94

EVWSI = -3490 + 40 = -3450 EVWOI = -3500.
EVSI = -3450 - (-3500) = \$50.
50>40 so conduct test.
If satisfactory report buy 2; if unsatisfactory
report buy 1

From the following tree we find EVWPI = -330. Then
EVPI = -3300 - (-3500) = 200.

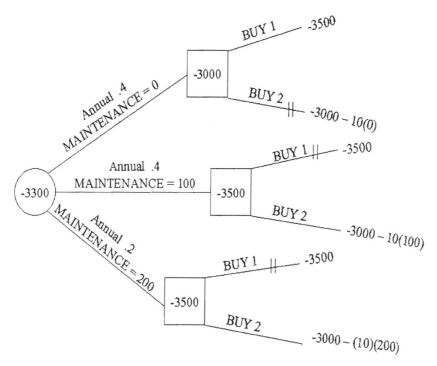

3. Let p = probability of stealing a base.
   Then we should steal if

1.194p+.243(1-p)>=.813 or p>=.599

95

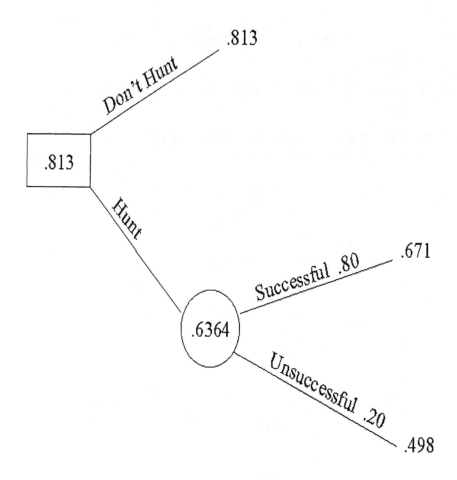

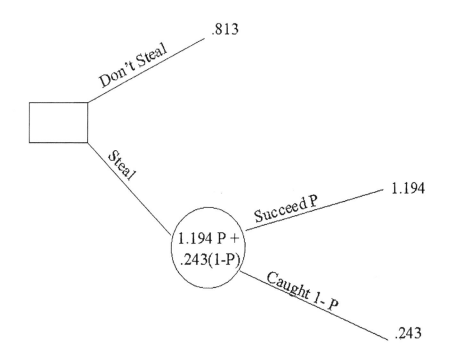

SECTION 13.5

1.

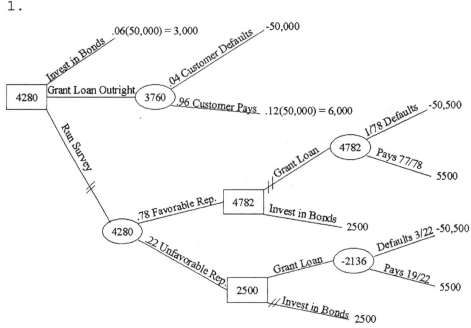

2. We are given p(earthquake) = .20 p(no earthquake)=.80
p(says quake|no quake) = .10, p(says no quake|no quake) = .9
p(says quake and quake) = .2(.95) = .19
p(says quake and no quake) = .8(.10) = .08
p(says quake) = .27 p(says no quake) = .73
p(says no quake and quake) = .2(.05)=.1
p(says no quake and no quake) = .8(.9) = .72

p(quake|says quake) = .19/.27 = 19/27
p(quake|says no quake) = .01/.73 = 1/73.
P(no quake|says quake) = .08/.27 = 8/27.
P(no quake|says no quake) =.72/.73 = 72/73

Thus we should hire geologist and if geologist predicts an earthquake build at Roy Rogers. If the geologist predicts no quake build at Diablo.

If the geologist were free EVWSI = -13.9 + 1 = -12.9. EVWOI = -14
EVSI = -12.90- (-14) 1.1 million

To determine EVPI look at 2nd tree. EVWPI = -12 so
EVPI = -12 -(-14) = 2 million dollars.

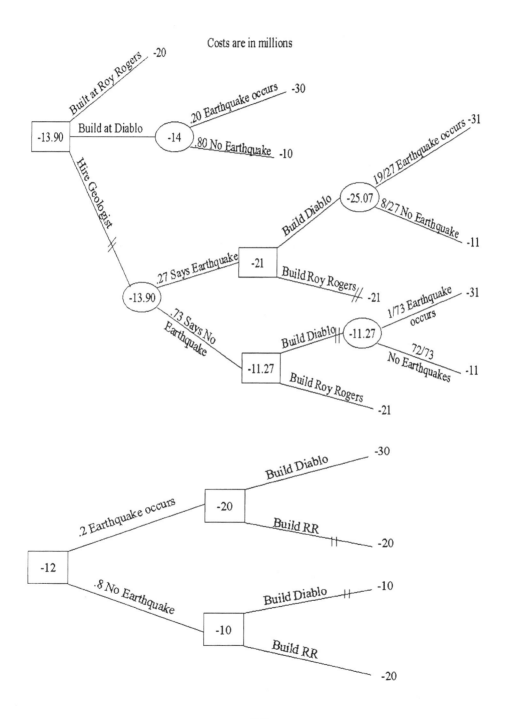

SECTION 13.6

1a. and 1b. Would need to have National decide between several lotteries to answer this question.

1c. Let x_1 = average cost per letter and x_2 = % of letters that are on-time. By assumption of mui we may write
(1) $u(x_1, x_2) = k_1 u_1(x_1) + k_2 u_2(x_2) + (1 - k_1 - k_2) u_1(x_1) u_2(x_2)$,
where $u_1(\$5) = 0$, $u_1(\$1) = 1$, $u_2(100\%) = 1$, $u_2(70\%) = 0$.
Also $u(\$1, 100\%) = 1$ and $u(\$5, 70\%) = 0$.
   From the first lottery and (1), we find that
$u(\$1, 70\%) = k_1 = .30$.
   From (1) and the second lottery we find that
$u(\$5, 100\%) = .50 = k_2$. Thus we find that (1) reduces to
$u(x_1, x_2) = .3u_1(x_1) + .5u_2(x_2) + .2u_1(x_1)u_2(x_2)$.
      Of course, we haven't been given enough information to determine $u_1(x_1)$ and $u_2(x_2)$.

Section 13.7

1. $A_{NORM} = \begin{bmatrix} 5/21 & 7/31 & 5/13 \\ 15/21 & 21/31 & 7/13 \\ 1/21 & 3/31 & 1/13 \end{bmatrix}$

$w_1 = (5/21 + 7/31 + 5/13)/3 = .283$
$w_2 = (15/21 + 21/31 + 7/13)/3 = .643$
$w_3 = (1/21 + 3/31 + 1/13)/3 = .074$

For teaching

$A_{NORM} = \begin{bmatrix} 4/5 & 4/5 \\ 1/5 & 1/5 \end{bmatrix}$

For research $\quad A_{NORM} = \begin{bmatrix} 1/4 & 1/4 \\ 3/4 & 3/4 \end{bmatrix}$

For service $\quad A_{NORM} = \begin{bmatrix} 6/7 & 6/7 \\ 1/7 & 1/7 \end{bmatrix}$

Professor 1 Score $= (4/5)w_1 + (1/4)w_2 + (6/7)w_3 = .45$
Professor 2 Score $= (1/5)w_1 + (3/4)w_2 + (1/7)w_3 = .55$
Thus professor 2 deserves a larger raise; this does not tell us how much larger, however.
To check for consistency we find

$Aw^T = [.87\ 2.01\ .22]$. Then

$(1/3)\{.87/.282 + 2.01/.643 + .22/.074\} = 3.06$ and
$CI = (3.06 - 3)/2 = .03$. $RI = .58$. Since $CI/RI < .1$, we have consistency of preferences.

2. $A_{NORM} = \begin{bmatrix} .1 & .077 & .118 \\ .4 & .308 & .294 \\ .5 & .615 & .588 \end{bmatrix}$

$w_1 = (.1 + .077 + .118)/3 = .098$
$w_2 = (.4 + .308 + .294)/3 = .334$
$w_3 = (.5 + .615 + .588)/3 = .568$

For cost we find that $A_{NORM} = \begin{bmatrix} .652 & .667 & .625 \\ .217 & .222 & .25 \\ .130 & .111 & .125 \end{bmatrix}$

Weights for Cost are easily found as follows

Computer 1 Cost = (.652 + .667 + .625)/3 = .648
Computer 2 Cost = (.217 + .222 + .25)/3 = .23
Computer 3 Cost = (.13 + .111 + .125)/3 = .122
Similarly, we find
Computer 1 User Weight = .154
Computer 2 User Weight = .640
Computer 3 User Weight = .206
Computer 1 Software Weight = .094
Computer 2 Software Weight = .168
Computer 3 Software Weight = .738
Computer 1 Score = .098(.648) + .334(.154) + .568(.094) = .169
Computer 2 Score = .098(.23) + .334(.64) + .568(.168) = .332
Computer 3 Score = .098(.122) + .334(.206) + .568(.738) = .500
Thus Computer 3 is preferred.
Checking for consistency, we find that
Aw^T = [2.95 1.011 1.726] and
{.295/.098 + 1.011/.334 + 1.726/.568}/3 = 3.025 and
CI = (3.025 - 3)/2 = .012. RI = .58, and CI/RI<.01, so again we have consistency of preferences.

Chapter 14 Solutions

Section 14.1

1.

			Row Min
	2	2	2
	1	3	1
Column Max	2	3	

Since max (row min) = min (column max) = 2, row choosing row 1 and column choosing column 1 is a saddle point. Value of the game is 2 units to the row player.

Section 14.2

Since the third column is dominated by either of the first two columns we need only solve the following game:

1	2
2	0

This game has no saddle point. Let x_1 = probability that row chooses row 1 and y_1 = probability that column chooses column 1.
Row player wants to choose x_1 to maximize

$\min(x_1 + 2(1 - x_1), 2x_1) = \min(2 - x_1, 2x_1)$. This maximum occurs where lines $y = 2 - x_1$ and $y = 2x_1$ intersect (at $x_1 = 2/3$), Value of game to row player is $2 - (2/3) = 4/3$.
   Column player chooses y_1 to minimize
$\max(y_1 + 2(1 - y_1), 2y_1) = \max(2 - y_1, 2y_1)$. The minimum occurs where $2 - y_1 = 2y_1$ or $y_1 = 2/3$. Thus value of game to row player is 4/3 and row player's optimal strategy is (2/3, 1/3) and the column player's optimal strategy is (2/3, 1/3, 0).

6. We have a constant-sum game which may be solved as if it is a zero-sum game with the following reward matrix:

		Firm 2	
		Low	High
Firm 1	Low	500	400
	High	300	600

We find that there is no saddle point so we let L = probability that firm 1's production level is low. Against Firm 2's Low production Firm 1's expected reward = 500L + 300(1 - L) = 200L + 300. Against Firm 2's high production strategy Firm 1's expected reward = 400L + 600(1 - L) = 600 - 200L. By the fundamental assumption of game theory, Firm 1 will receive min[600 - 200L, 200L + 300]. This expression is maximized when 600 - 200L = 200L + 300 or L = 3/4. This yields value of game = $450 to Firm 1.

Let L' = probability that firm 2 chooses low production. Then Firm 2 wishes to minimize Firm 1's expected reward. Against Firm 1's low production level Firm 1 gains 500L' + 400(1 - L') = 400 +100L'. Against Firm 1's high production strategy, Firm 1's expected reward is 300L' + 600(1 - L') = 600 - 300L'. By the fundamental assumption of game theory, Firm 2's choice of L' will result in an expected reward to Firm 1 = max [100L' + 400, 600 - 300L']. This maximum occurs for L' = .5. Thus Firm 1 should choose low production level 3/4 of time while Firm 2 should choose a low production level only 1/2 the time.

Section 14.3

1a.

	Foxhole 1	Foxhole 2	Foxhole 3	Foxhole 4	Foxhole 5
Row 1 A	1	1	0	0	0
Row 2 B	0	1	1	0	0
Row 3 C	0	0	1	1	0
Row 4 D	0	0	0	1	1

1b. Column 2 is dominated by column 1 while column 4 is dominated by column 5.

1c. Suppose gunner chooses strategy which shoots at A, C, and D, 1/3 of the time. Then (assuming that the soldier plays his optimal strategy)no matter where the soldier hides, there is a 1/3 chance that the soldier will be killed. Thus the value to the gunner is 1/3.

1d. If the soldier chooses the given non-optimal strategy, and the gunner always fires at A, the gunner will earn an expected reward of 1/2>1/3.

1e. Row 1 = A,... Row 4 = D

Gunner's LP   Soldier's LP

max v   min w

st. $v \leq x_1$ st $w \geq y_1 + y_2$
$v \leq x_1 + x_2$ $w \geq y_2 + y_3$
$v \leq x_2 + x_3$ $w \geq y_3 + y_4$
$v \leq x_4$, $v \leq x_3 + x_4$ $w \geq y_4 + y_5$
$x_1 + x_2 + x_3 + x_4 = 1$ $y_1 + y_2 + y_3 + y_4 + y_5 = 1$
$x_1, x_2, x_3, x_4 \geq 0$ $y_1, y_2, y_3, y_4, y_5 \geq 0$

For soldier $y_1 = y_3 = y_5 = w = 1/3$ is feasible. For gunner $x_1 = x_3 = x_4 = v = 1/3$ is also feasible. Since these solutions have v = w, they are both optimal strategies.

Section 14.4

3. Rewards are in millions. We assume that each borough will support its own bond issue (this dominates not supporting your own bond issue). Then the reward matrix is as follows:

<table>
<tr><td></td><td colspan="2">Brooklyn Strategies</td></tr>
<tr><td>Manhattan</td><td>Support Manhattan</td><td>Oppose Manhattan</td></tr>
<tr><td>Support Brooklyn</td><td>(8,8)</td><td>(-1,9)</td></tr>
<tr><td>Oppose Brooklyn</td><td>(9,-1)</td><td>(0,0)</td></tr>
</table>

This is a Prisoner's dilemma game with (0, 0) being an equilibrium point. "Oppose" is the non-cooperative action while "Support" is the cooperative action.

Section 14.7

Let (x_1, x_2, x_3, x_4) be a point in the core. Then x_1, x_2, x_3, and x_4 must satisfy

(1) $x_1 + x_2 + x_3 \geq 75$ (2) $x_1 + x_2 + x_4 \geq 75$
(3) $x_1 + x_3 + x_4 \geq 75$ (4) $x_2 + x_3 + x_4 \geq 75$
(5) $x_1 + x_2 + x_3 + x_4 \geq 100$ (6) $x_3 + x_4 \geq 60$

$$x_i \geq 0.$$

Since (x_1, x_2, x_3, x_4) must be an imputation we also require that
(7) $x_1 + x_2 + x_3 + x_4 = 100$. Adding (1)-(4) yields $3(x_1 + x_2 + x_3 + x_4) \geq 300$ or $x_1 + x_2 + x_3 + x_4 \geq 100$. By (7) we now know that (1)- (4) must all hold with equality. Thus any point in the core must satisfy $x_1 = x_2 = x_3 = x_4 = 25$. This, however, violates (6), so the core is empty.

2. Inequality (6) in solution to Problem 1 is now $x_3 + x_4 \geq 50$.
Since (25, 25, 25, 25) satisfies this inequality the core is the point (25, 25, 25, 25).

Chapter 15 Solutions

1a, 1b and 1c D = 4,000 gallons/month h = $.025/gallon-month K = $50

$$q = \sqrt{\frac{2(50)(4,000)}{.025}} = 4000 \text{ gallons}$$

Orders/Month = D/q = 4000/4000 = 1 order/month = 12 orders/year. There will be one month between orders.

1d. Demand is usually random and varies with the day of the week. If this is the case then the EOQ model should not be used (see Section 16.6)

1e. If Leadtime = 2 weeks, then Leadtime demand<EOQ and order should be placed when (2/52)(48,000) = 1846 gallons are in stock. If Leadtime = 10 weeks, then Leadtime demand>EOQ. . Assume that an order arrives at time T = 0. Then an order must have been placed at T = -10. We know that an order is placed every 52/12 = 13/3 weeks. Thus an order was placed at T = -17/3, T = -4/3, T = 3, etc. Since at T = 0 we will have 4000 gallons in stock, we know that at the reorder point T = 3 (and any other reorder point) we will have

4000 - (3/52)48,000 = 1231 gallons

in stock.

2. K = (1/4)10 = $2.50 D = $10,000/year h = $.10/dollar-year

2a and 2b q = $\sqrt{\frac{2(2.50)10,000}{.10}}$ = $707.11 should be withdrawn on each trip to the bank.

$$\frac{\text{Trips}}{\text{Year}} = \frac{10{,}000}{707.11} = 14.14 \text{ trips/year}$$

2c. Since $D/q = (hD/2K)^{1/2}$ an increase in D will increase D/q = Trips to bank/year.

2d. This increases h which causes q to decrease. This will increase the number of trips to the bank each year.

2e. Increasing the number of tellers should reduce the time we spend waiting in line. This reduces K which leads to a decrease
in q. Since q decreases, we will go to the bank more often.

SECTION 15.3

1. D = 960 boxes/year K = $20
For a $9.70 price h = .2(9.7) = 1.94 so in this case

$$EOQ = \sqrt{\frac{2(20)(960)}{1.94}} = 140.69$$

Hence for a $9.70 price q = 500 is the best we can do.
For a $9.80 price h = .2(9.80) = 1.96 so in this case

$$EOQ = \sqrt{\frac{2(20)(960)}{1.96}} = 139.97$$

Since EOQ<300 the best we can do for a $9.80 price is to choose q = 300.
For a $10 price h = .2(10) = 2 and

$$EOQ = \sqrt{\frac{2(20)960}{2}} = 138.56$$

108

Since 138.56<300, ordering 138.56 is admissible and the best we can do for a $10 price is choose q = 138.56.

We now compare the total annual cost for q = 138.56, q = 300, and q = 500.

Q	Ann. Purchase Cost	Ann. Holding Cost	Ann. Order Cost
138.56	960(10) = $9600	(1/2)2(138.56) = $138.56	20(960)/138.56 = $138.56
300	960(9.8) = $9408	(1/2(1.96)(300 = $294	(20(960)/300 = $64
500	960(9.7) = $9312	(1/2)(1.94)(500) = $485	20(960)/500 = $38.40

Then TC(138.56) = $9877.12, TC(300) = $9766 and TC(500) = $9835.40.
Hence q = 300 minimizes total annual cost.

SECTION 15.4

1. Optimal Run Size = $EOQ[r/(r-D)]^{1/2}$. Since $r/(r-D)>1$, we see that Optimal Run Size>EOQ . Intuitively, the optimal run size exceeds the EOQ because the `gradual' production that occurs in the rate model reduces the holding costs associated with a given value of q. This allows us to choose a higher run size than in the EOQ model.

SECTION 15.5

1. Since $(h + s)/s>1$, $q^*>EOQ$. Since $s/(h + s)<1$, M*<EOQ.

2. D = 500 cars/year ,s = $20,000/car-year, K = $10,000,
h = $5,000/car-year

$$q^* = \sqrt{\frac{2(10,000)(500)(25,000)}{5,000(20,000)}} = 50 \text{ cars}$$

$$M^* = \sqrt{\frac{2(10,000)(5,000)(20,000)}{(5,000)(25,000)}} = 40 \text{ cars}$$

Thus the maximum shortage = $q^* - M^* = 10$ cars

Section 15.6

1. $\bar{d} = 600/4 = 150$

Est Var D = $1/4\{100^2 + 50^2 + 150^2 + 300^2\} - (150)^2 =$
31,250-22,500
= 8750

and

$$VC = \frac{8750}{(150)^2} = .389$$

Since VC>.20 demand is too lumpy to justify the use of an EOQ model.

Section 15.7

1. See file S15_7_1.xls.

110

Chapter 16 Solutions

SECTION 16.2

1a. q = 6

1b. Since $E(2) - E(1) < 0$ and $E(3) - E(2) > 0$ marginal analysis indicates that q = 2 is optimal.

1c. Marginal analysis fails in this example because $E(q)$ is not a convex function of q. Thus while q = 2 is a local minimum it need not minimize q over all values of q.

SECTION 16.3

1. For d≤q Total Cost = 10,000q - 15,000d - 9000(q - d)

   = 1,000q - 6000d

   For d≥q + 1 Total Cost = 10,000q - 15,000d + 12,000(d - q)

   = -2000q - 3000d

   Thus $c_0 = 1,000$ and $c_u = 2,000$. Hence we should order q* where q* is the smallest value of q satisfying

$$P(D \leq q^*) \geq \frac{2000}{3000} = \frac{2}{3} = .667$$

Since $P(D \leq 30) = .60$ and $P(D \leq 35) = .80$ 35 cars should be ordered in August.

2. For d≤q Total Cost = 15q - 30d.

   For d≥q + 1 Total Cost = 15q - 30q = -15q.

Thus $c_0 = 15$ and $c_u = 15$. Hence we should order q* papers where q* is the smallest value of q satisfying

$$P(D \leq q^*) \geq \frac{15}{15 + 15} = .50$$

Since $P(D \leq 70) = .45$ and $P(D \leq 90) = .70, 90$ papers should be ordered.

SECTION 16.4

1a. In all likelihood each ticketed customer has a fixed probability, say .05, of not boarding the flight. If this is the case, then the number of no-shows would have a mean of .05q, which depends on the number of tickets sold.

1b. No, because the newsboy problem assumes that f(d) is given and does not depend on q.

2. Let d=Amount of funds actually needed and q=Amount of funds borrowed from bank
If $d \leq q$ Total Cost = .10q
If $d \geq q$ Total cost = .10q + .25(d - q) = .25d - .15q
Thus $c_0 = .10$ and $c_u = .15$. q* dollars should be borrowed where
q* satisfies

$$P(D \leq q*) = \frac{.15}{.15 + .10} = .60$$

Standardizing we obtain

$$P(Z \leq \frac{q* - 700,000}{300,000}) = .60$$

Since $F(.25) = .60$, $\frac{q* - 700,000}{300,000} = .25$

Then q*=$775,000.

3. If $d \leq q$ cost is 10q - 25d while if $d \geq q$ cost is 10q - 25q. Thus
$c_0 = \$10$ and $c_u = \$15$ and we should order q* trees, where

$$P(D \leq q*) = 15/(10 + 15) = .60 .$$

Standardizing yields

$$P(Z \leq \frac{q^* - 100}{30}) = .60$$

Since $F(.25) = .60$, we find that $\frac{q^* - 100}{30} = .25$ or

$q^* = 107.5$ trees

4a. d = demand q = hot dogs ordered
$c(d,q) = -1.5d + 1.2q - (q-d)$ for $d \leq q$, $c_o = .2$
$c(d,q) = -1.5q + 1.2q$ for $d \geq q+1$, $c_u = .3$
$F(q^*) = .3/(.3+.2) = .6$. $P(D \leq q^*) = 60.$
Standardizing yields $q^* = 40 + 10(.25) = 42.5$ hot dogs.

4b. $P(D \leq 52) = P(Z \leq 1.2) = .885$

5. q = capacity(in thousands of sets) d = annual demand(in thousands of sets) Costs are per year.
$c(d,q) = 100,000q - 250,000d$ $d \leq q$, $c_o = 100,000$
$c(d,q) = 100,000q - 250,000q$, $d > q+1$, $c_u = 150,000$.
Choose capacity level q^* so that
$P(D \leq q^*) = 150,000/(100,000+250,000) = .6.$
Standardizing yields $q^* = 6000 + 2000(.25) = 6500$ sets

6. Let **D** be the random variable for the number of orders received in a day and **H** be the random variable for the number of employee hours needed during a day.
 Let d=Actual Number of orders received during a day
      q=Actual Number of fulltime employees
and h=Actual Number of hours of employee labor required during a day. Note that both d and h are values of random variables.
For $h \leq 8q$ Total Cost = 80q and for
For $h \geq 8q$ Total Cost = $80q + (h - 8q)15$.
We may rewrite the above as
  For $(h/8) \leq q$ Total Cost = 80q
  For $(h/8) \geq q$ Total Cost = $-40q + 120(h/8)$
We now have a newsboy problem with the random `demand' following the random variable **H**/8

Also $c_0=80$ and $c_u=40$. Since each worker can handle 50/8=6.25
orders per hour it must be true that h=d/6.25.
Then h/8=d/50. In terms of **H** and **D** we may state that
H/8=**D**/50.Hence **H**/8is normally distributed with
E(**H**/8) = 2000/50 = 40 and (Std. Dev. **H**/8) = 500/50 = 10.
We now know that the optimal number of employees
(call it q*) must satisfy
$$P(\mathbf{H}/8 \leq q*) = 40/(80+40) = .333$$
Standardizing with respect to the random variable
H/8 we obtain

$$P(Z \leq \frac{q* - 40}{10}) = .333. \text{ Hence } F(\frac{q* - 40}{10}) = .333$$

Since F(-.43)=.333 we have that

$$\frac{q* - 40}{10} = -.43 \quad \text{ or } q*=35.7.$$

Therefore 35 or 36 full time employees should be
working during the Christmas season.

7. For d≤q cost is given by $c_1(d, q) = c_oq + cd + k$
   For d≥q cost is given by $c_2(d, q) = -c_uq + c'd + k'$
 Also note that for any q , $c_1(q,q) = c_2(q,q)$
Letting E(q) be total expected cost we find that

$$E(q) = \int_0^q c_oqf(t)dt + \int_q^\infty -c_uqf(t)dt$$

$$+\int_0^q (\overline{ct} + \overline{k})f(t)dt + \int_q^\infty (c't + k')f(t)dt$$

Leibnitz's rule and $c_1(q, q) = c_2(q, q)$ yields
(F(q) = P(**D**≤q))
$E'(q) = c_oF(q) - c_u(1 - F(q)) = 0$ for q* given by
$F(q*) = c_u/(c_o + c_u)$.
Since $E''(q*) = (c_o + c_u)f(q*) > 0$, we know that q*
yields a minimum, and not a maximum.

SECTION 16.5

1. Let x = Location of a fire and s = location of the fire station.

If s≤x then the distance between the fire station and the fire is x - s while if s≥x the distance between the fire station and the fire is s - x. As a function of the location of the fire station, the expected distance between the fire station and a fire is given by

$$D(s) = \int_0^s (s - x)2x\,dx + \int_s^1 (x - s)2x\,dx$$

$$= [x^2 s - 2x^3/3]_0^s + [2x^3/3 - x^2 s]_0^1$$

$$= 2s^3/3 - s + 2/3$$

$D'(s) = 2s^2 - 1 = 0$ for $s* = 2^{1/2}/2$. Since $D''(s) = 4s \geq 0$,
$D(s)$ is a convex function and $s* = 2^{1/2}/2$ does minimize $D(s)$.

SECTION 16.6

1. K = $20 , E(**D**) =1040 , Std. Dev. **D** = 43.26, L = 1/52 year,
h = $20/pint-year, c_B = $50/pint
E(**X**)=1040/52=20 and Std. Dev. **X**=43.26/$(52)^{1/2}$ = 6.

$$EOQ = \sqrt{\frac{2(20)(1040)}{20}} = 45.61$$

From (13)

$$P(\mathbf{X} \geq r) = \frac{20(45.61)}{50(1040)} = .018$$

115

Standardizing we obtain

$$P(Z \geq \frac{r - 20}{6}) = .018$$

Since $F(2.10) = .982$ we find that $r = 20 + 6(2.10) = 32.60$ and the safety stock level is $32.60 - 20 = 12.60$.

Of course, in this problem we are making the unrealistic assumption that blood does not spoil!

We set $s = r = 32.6$ and $S = r + EOQ = 32.6 + 45.61 = 78.21$.

SECTION 16.7

1. From Problem 1 in Section 16.6 , $q = 45.61$, $E(X) = 20$,
Std. Dev. $X = 6$
For 80% Service Level we need

$$NL(\frac{r-20}{6}) = \frac{45.61(.2)}{6} = 1.52$$

Since $NL(-1.49) = 1.52$ we have that $\frac{r-20}{6} = -1.49$ or

$r = 20 - 6(1.49) = 11.06$
For a 90% Service Level we need

$$NL(\frac{r-20}{6}) = \frac{(45.61)(.1)}{6} = .76$$

Since $NL(-.59) = .76$

$$\frac{r-20}{6} = -.59 \text{ or } r = 20 - 6(.59) = 16.46$$

For a 95% Service Level

$$NL(\frac{r-20}{6}) = \frac{45.61(.05)}{6} = .38$$

Since $NL(.04) = .38$, $\dfrac{r-20}{6} = .04$ and $r = 20 + (.04)6 = 20.24$

For a 99% Service Level

$$NL\left(\dfrac{r-20}{6}\right) = \dfrac{45.61(.01)}{6} = .076$$

Since $NL(1.05) = .076$ we have that $r = 20+6(1.05) = 26.30$

To determine the reorder point that will ensure at most 0.5 stockouts per year we use (19). We want

$$P(\mathbf{X} \geq r) = \dfrac{(0.5)(45.61)}{1040} = .022$$

Standardizing yields

$$P\left(\mathbf{Z} \geq \dfrac{r - 20}{6}\right) = .022. \text{ Since } F(2.01) = .978,$$

$$\dfrac{r - 20}{6} = 2.01 \text{ and } r = 32.06.$$

2. $q = 107.50$, $E(\mathbf{X}) = 40$ Std. Dev. $\mathbf{X} = 10$
For 80% Service Level

$$NL\left(\dfrac{r-40}{10}\right) = \dfrac{(107.50).20}{10} = 2.15$$

Since $NL(-2.14) = 2.15$ $r = 40 - 2.14(10) = 18.6$
For 90% Service Level

$$NL\left(\dfrac{r-40}{10}\right) = \dfrac{(107.50)(.10)}{10} = 1.075$$

Since $NL(-.99) = 1.075$ we find that $r = 40 - .99(10) = 30.1$
For 95% Service Level

$$NL\left(\dfrac{r-40}{10}\right) = \dfrac{(107.50)(.05)}{10} = .538$$

Since $NL(-.25) = .536$ we find that $r = 40-.25(10) = 37.5$
For 99% Service level

$$NL\left(\frac{r-40}{10}\right) = \frac{(107.5)(.01)}{10} = .108$$

Since NL(.86) = .108 we find that r = 40 + .86(10) = 48.60

We now use (19) to find the reorder point which yields an average of 2 stockouts per year.

$$P(\mathbf{X} \geq r) = \frac{2(107.50)}{1040} = .207$$

Standardizing yields $P\left(\mathbf{Z} \geq \dfrac{r - 40}{10}\right) = .207$. Since F(.82)

= .794 is closest to .793

$$\frac{r - 40}{10} = .82 \text{ or } r = 48.2$$

3a. Expected Shortages/Cycle = 5(1/4) = 1.25
Thus Expected Demand Met on Time/Cycle = 100 - 1.25 = 98.75
Since there are 1000/100 = 10 cycles per year

$$\text{Service Level} = \frac{987.50}{1,000} = .9875$$

3b. Choose the smallest order size that will result in an average of 5 or fewer shortages per cycle. (this yields 50 stockouts per year which is a 95% service level.)
r = 10 yields 1/4(5) + 1/4(10) + 1/12(15) + 1/4(20)
= 10 stockouts/cycle.
r = 15 leads to (1/4)5 + (1/12)10 + (1/4)15 = 35/6 shortages/cycle
r = 20 leads to (1/12)5 + (1/4)10 = 35/12 shortages/cycle
Hence r=20 is the smallest reorder point that will

provide a 95% service level.

3c. Choose r to be the smallest number satisfying
$$P(\mathbf{X}>r) = \frac{2(100)}{1000} \leq .20$$
$P(\mathbf{X}>29) = 1/4$ and $P(\mathbf{X}>30) = 0$, so $r = 30$ is required to ensure that at an average of at most two stockouts per year will occur.

4. $NL\left(\dfrac{r-1000}{200}\right) = 780(1 - .90)/200 = .39.$

$NL(.02) = .39.$

$q = [2(50)(365)(100/6]^{1/2} = 780.$ We find $r = 1000 + .02(200) = 1004.$ Safety Stock $= 1004 - 1000 = 4.$ Annual Safety Stock holding Cost $= 4h = \$24.$

SECTION 16.8

1. EOQ $= [2(10,000)(530)/5]^{1/2} = 1456$ R $= 1456/10,000 = .1456$
years. R + L $= .23$ years. D_{L+R} is $N(2300, 384^2)$. $P(D_{L+R} \geq S) = (.1456)5/100 = .007.$ Since $P(Z \geq 2.45) = .007$ S $= 2300 + 384(2.45) = 3241.$

SECTION 16.9

1. Type A Items: 1 and 2 Type B Items:3-6 Type C Items:7-10. See graph. Note that we have slightly more Type B items than the text indicates, but that's ok.

Section 16.10

1a. See graph

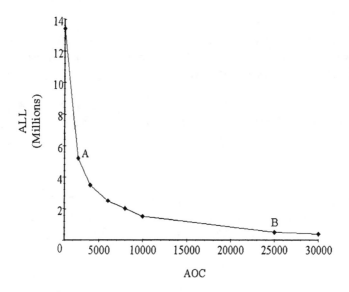

1b. we have AOC = 2(500+800) = $2600 and AII = .5*(5000)(2000) + .5*(10,000)(250) = $6,250,000. Point A has AOC of $2600 and lower AII, so clearly the current policy can be improved upon.

1c. All points on curve to left of B are permissible.

Chapter 17 Solutions

1. Letting the state for a day be the day's weather yields a two state Markov Chain with the following transition matrix:

	Sunny	Cloudy
Sunny	.90	.10
Cloudy	.20	.80

2.

$$p = \begin{array}{c} \\ 0 \\ 1 \\ 2 \\ 3 \\ 4 \end{array} \begin{array}{c} \begin{array}{ccccc} 0 & 1 & 2 & 3 & 4 \end{array} \\ \left[\begin{array}{ccccc} 0 & 0 & 1/3 & 1/3 & 1/3 \\ 0 & 0 & 1/3 & 1/3 & 1/3 \\ 1/3 & 1/3 & 1/3 & 0 & 0 \\ 0 & 1/3 & 1/3 & 1/3 & 0 \\ 0 & 0 & 1/3 & 1/3 & 1/3 \end{array}\right] \end{array}$$

1. U=Urban S=Suburban R=Rural. Then

$$p = \begin{array}{c} \\ U \\ S \\ R \end{array} \begin{array}{c} \begin{array}{ccc} U & S & R \end{array} \\ \left[\begin{array}{ccc} .80 & .15 & .05 \\ .06 & .90 & .04 \\ .04 & .06 & .90 \end{array}\right] \end{array}$$

$$\text{a.} \; P_{UU}(2) = [.80 \;.15 \;.05] \begin{bmatrix} .80 \\ .06 \\ .04 \end{bmatrix} = .651$$

$$P_{US}(2) = [.80 \;.15 \;.05] \begin{bmatrix} .15 \\ .90 \\ .06 \end{bmatrix} = .258$$

$$P_{UR}(2) = [.80 \ .15 \ .05] \begin{bmatrix} .05 \\ .04 \\ .90 \end{bmatrix} = .091$$

Note that these probabilities sum to one; after all, after two years an urban dweller must be somewhere.

b. $q = [.40 \ .35 \ .25]$. We seek

$$q(\text{Column 1 of } P^2) = [.40 \ .35 \ .25] \begin{bmatrix} .651 \\ .104 \\ .072 \end{bmatrix}$$

$$= .315$$

c. The moving tendencies of Americans change over time. After all, people used to move into urban areas, and now people are moving out of urban areas. Thus we have, in all probability, a non-stationary Markov Chain.

SECTION 17.4

1. Period of states 1, 2, and 3 is 2. To return to any state (other than 0 or 4) the number of wins player has had must equal number of losses. Thus gambler can only return to original state after an even number of plays.

2. Yes, all states communicate and no states are periodic.

SECTION 17.5

1. After replacing the third steady state equation by $\pi_1 + \pi_2 + \pi_3 = 1$ we obtain the following system (1 = U, 2 = S, 3 = R).
$\pi_1 = .80\pi_1 + .06\pi_2 + .04\pi_3$
$\pi_2 = .15\pi_1 + .90\pi_2 + .06\pi_3$

$\pi_1 + \pi_2 + \pi_3 = 1$
Solving these equations we find that $\pi_1 = 38/183$, $\pi_2 = 90/183$, and $\pi_3 = 55/183$. Thus about 49% of the families will eventually live in suburban areas; about 21% in urban areas; and about 30% in rural areas.

3a. $\pi_1 = 2\pi_1/3 + .5\pi_2$ and $\pi_1 + \pi_2 = 1$. Thus $\pi_1 = .6$ and $\pi_2 = .4$.

3b. $\pi_1 = .8\pi_1 + .8\pi_3$, $\pi_3 = .8\pi_2$, and $\pi_1 + \pi_2 + \pi_3 = 1$. Solving these equations we find that $\pi_1 = 16/25$, $\pi_2 = 1/5$, and $\pi_3 = 4/25$.

3c. $m_{11} = 1/\pi_1 = 1/.64 = 1.56$, $m_{12} = 1 + .8m_{13}$, $m_{13} = 1 + .8m_{13} +$
$.2m_{23}$, $m_{21} = 1 + .2m_{21} + .8m_{31}$, $m_{22} = 1/\pi_2 = 1/.2 = 5$, $m_{23} = 1 +$

$.2m_{23}$, $m_{31} = 1 + .2m_{21}$, , $m_{32} = 1 + .8m_{12}$, $m_{33} = 1/\pi_3 = 1/.16 = 6.25$. Solving these equations we find that $m_{11} =$
1.56, $m_{12} = 6$, $m_{13} = 6.25$, $m_{23} = 1.25$, $m_{21} = 2.81$, $m_{31} = 1.56$,
$m_{22} = 5$, $m_{32} = 5.8$, $m_{33} = 6.25$.

8a. Let the state during a period equal the number of balls in Container 1. Then the transition probability matrix is

	0	1	2	3
0	0	1	0	0
1	1/3	0	2/3	0
2	0	2/3	0	1/3
3	0	0	1	0

Find the steady state probabilities by solving
$\pi_0 = \pi_1/3$, $\pi_1 = \pi_0 + 2\pi_2/3$, $\pi_2 = 2\pi_1/3 + \pi_3$, $\pi_0 + \pi_1 + \pi_2 + \pi_3 = 1$.
We obtain $\pi_0 = 1/8$, $\pi_1 = 3/8$, $\pi_2 = 3/8$, $\pi_3 = 1/8$.

8b. $m_{0,0} = 1/\pi_0 = 8$.

1.

$$Q = \begin{array}{c} \\ Fr \\ So \\ Jr \\ Sr \end{array} \begin{array}{cccc} Fr & So & Jr & Sr \\ \begin{bmatrix} .1 & .8 & 0 & 0 \\ 0 & .1 & .85 & 0 \\ 0 & 0 & .15 & .80 \\ 0 & 0 & 0 & .10 \end{bmatrix} \end{array}$$

$$R = \begin{array}{c} \\ Fr \\ So \\ Jr \\ Sr \end{array} \begin{array}{cc} Q & G \\ \begin{bmatrix} .1 & 0 \\ .05 & 0 \\ .05 & 0 \\ .05 & .85 \end{bmatrix} \end{array}$$

$$I - Q = \begin{bmatrix} .9 & -.8 & 0 & 0 \\ 0 & .9 & -.85 & 0 \\ 0 & 0 & .85 & -.80 \\ 0 & 0 & 0 & .10 \end{bmatrix}$$

$$(I-Q)^{-1} = \begin{bmatrix} 1.11 & .99 & .99 & .88 \\ 0 & 1.11 & 1.11 & .99 \\ 0 & 0 & 1.18 & 1.05 \\ 0 & 0 & 0 & 1.11 \end{bmatrix}$$

1a. Exp. Number of Years as Freshman = $(I-Q)^{-1}_{11}$ = 1.11

Exp. Number of Years as Sophomore = $(I-Q)^{-1}_{12}$ = .99

Exp. Number of Years as Junior = $(I-Q)^{-1}_{13}$ = .99

Exp. Number of Years as Senior = $(I-Q)^{-1}_{14}$ = .88

so Exp. Number of Years an Entering Freshman spends at I.U. is
1.11 + .99 + .99 + .88 = 3.97. Note that this is <4 yrs. because many entering freshman quit before their senior year.

1b. $\{(I-Q)^{-1}R\}_{12} = [1.11 \quad .99 \quad .99 \quad .88] \begin{bmatrix} 0 \\ 0 \\ 0 \\ .85 \end{bmatrix} = .748$

2.

	New Sub.	1 Yr. Sub.	>=2 Yr. Sub	Cancelled
New Sub.	0	.80	0	.20
1 Yr. Sub	0	0	.90	.10
>=2 Yr. Sub	0	0	.96	.04
Cancelled	0	0	0	1

$$Q = \begin{bmatrix} 0 & .80 & 0 \\ 0 & 0 & .90 \\ 0 & 0 & .96 \end{bmatrix} \qquad R = \begin{bmatrix} .4 & .1 & 0 & 0 \\ 0 & 0 & .2 & .5 \end{bmatrix}$$

$$(I-Q)^{-1} = \begin{bmatrix} 1 & .80 & 18 \\ 0 & 1 & 22.5 \\ 0 & 0 & 25 \end{bmatrix}$$

We are starting in state 1 (new Subscriber) Thus expected number of years that person remains a subscriber is 1 + .80 + 18 = 19.80 years.

10. Without the GRP
$$Q = \begin{matrix} .991 & .003 \\ .025 & .969 \end{matrix} \qquad I-Q = \begin{bmatrix} .009 & -.003 \\ -.025 & .031 \end{bmatrix}$$

Approximately $\qquad (I-Q)^{-1} = \begin{bmatrix} 152 & 15 \\ 123 & 44 \end{bmatrix}$

125

Thus Without GRP (remember patient starts in hospital)
Exp. Number of months Patient Spends in Hospital $= (I-Q)^{-1}{}_{11} = 152$
Exp. Number of Months Patient Spends in Home $= (I-Q)^{-1}{}_{12} = 15$
Thus Exp. Cost per Patient $= 655(152) + 226(15) = \$102,950$.
With the GRP

$$Q = \begin{bmatrix} .854 & .028 & .112 & 0 \\ .013 & .978 & 0 & .003 \\ .025 & 0 & .969 & 0 \\ 0 & .025 & 0 & .969 \end{bmatrix}$$

$$I - Q = \begin{bmatrix} .146 & -.028 & -.112 & 0 \\ -.013 & .022 & 0 & -.003 \\ -.025 & 0 & .031 & 0 \\ 0 & -.025 & 0 & .031 \end{bmatrix}$$

Approximately,

$$(I-Q)^{-1} = \begin{bmatrix} 27 & 39 & 97 & 4 \\ 18 & 77 & 65 & 7 \\ 22 & 31 & 111 & 3 \\ 14 & 62 & 52 & 38 \end{bmatrix}$$

Since a patient starts in the hospital (state 2)
Exp. Number of Months Patient Spends in GRP = 18
Exp. Number of Months Patient Spends in Hospital = 77
Exp. Number of Months Patient Spends in Home=65 + 7 = 72
Hence with the GRP. Exp. Cost Per Patient = 680(18) + 655(77) + 226(72) = \$78,947.
Thus the GRP does save money (not to mention the improvement in the patient's quality of life: less time in hospital, more time in home.

10b. Without GRP average patient spends 152 months in hospital; with GRP average patient spends 77 months in the hospital.

SECTION 17.7 SOLUTIONS

1. Steady State Equations are
Number entering yr. i of college = Number leaving yr. i of college

$$7,000 = (.8+.1)N_1$$
$$500+.8N_1 = (.85+.05)N_2$$
$$500+.85N_2 = (.80+.05)N_3$$
$$0 +.80N_3 = (.05+.85)N_4$$

Solving we find that $N_1 = 7777.78$, $N_2 = 7469.14$, $N_3 = 8057.37$
$N_4 = 7162.10$. Thus the juniors will be the most numerous class.

2. This changes the third steady state equation to
$.03N_2 = .03N_3$

which forces $N_2 = N_3$. Since the number of working adults will equal the number of retirees each working adult must, in effect, pay the pension for one retiree. Thus annual contribution by each working adult must be $5,000.

Chapter 18 Solutions

1. If I can force it to be my opponent's turn with 1 match left, I will win. Working backwards, if I can force my opponent's turn to occur with 6, 11, 16, 21, 26, 31 or 36 matches on the table I will win. Thus I should pick up 40- 36 = 4 matches on the first turn and on each successive turn pick up (5 - # of matches my opponent has picked up on his last turn).

1. Define $f_t(i)$ to be the shortest path from node i to node 10
given that node is a stage t node.
  Define $x_t(i)$ to be the endpoint of the arc that should be
chosen if we are in node i.
$f_4(8) = 3 \quad x_4(8) = 10$
$f_4(9) = 4 \quad x_4(9) = 10$

$$f_3(5) = \min \begin{cases} 1 + f_4(8) = 4* & x_3(5) = 8 \\ \\ 3 + f_4(9) = 7 \end{cases}$$

$$f_3(6) = \min \begin{cases} 6 + f_4(8) = 9 \\ \\ 3 + f_4(9) = 7* & x_3(6) = 9 \end{cases}$$

$$f_3(7) = \min \begin{cases} 3 + f_4(8) = 6* & x_3(6) = 8 \\ \\ 3 + f_4(9) = 7 \end{cases}$$

$$f_2(2) = \min \begin{cases} 7 + f_3(5) = 11* \\ \\ 4 + f_3(6) = 11* & x_2(2) = 5 \text{ or } 6 \\ \\ 6 + f_3(7) = 12 \end{cases}$$

$$f_2(3) = \min \begin{cases} 3 + f_3(5) = 7^* & x_2(3) = 5 \\ 2 + f_3(6) = 9 \\ 4 + f_3(7) = 10 \end{cases}$$

$$f_2(4) = \min \begin{cases} 4 + f_3(5) = 8^* \\ 1 + f_3(6) = 8^* & x_2(4) = 5 \text{ or } 6 \\ 5 + f_3(7) = 11 \end{cases}$$

$$f_1(1) = \min \begin{cases} 2 + f_2(2) = 13 \\ 4 + f_2(3) = 11^* & x_2(1) = 3 \text{ or } 4 \\ 3 + f_2(4) = 11^* \end{cases}$$

From this analysis we see that 1-3-5-8-10, 1-4-6-9-10, or 1-4-5-8-10 are all shortest paths from node 1 to node 10 (each of these paths has length 11).

SECTION 18.3

2. Let $f_t(i)$ be the minimum cost incurred during months $t, t + 1, ..3$ if the inventory at the beginning of month t is i.
Note that it is clearly suboptimal to produce more than the total demand of 800 units. This means that our inventory at the beginning of month 3 can't exceed 800 - (200 + 300) = 300 units.
Thus we need only compute $f_3(i)$ for $i = 0, 100, 200$ and 300. As in the text, during month 3 we simply produce enough to meet month 3 demand from current inventory and production. Let $x_t(i)$ be the quantity that should be produced during month t in order to attain $f_t(i)$. Then

$f_3(300) = 0 \quad x_3(300) = 0$
$f_3(200) = 250 + 12(100) = 1450 \quad x_3(200) = 100$
$f_3(100) = 250 + 12(200) = 2,650 \quad x_3(100) = 200$
$f_3(0) = 250 + 12(300) = 3,850 \quad x_3(0) = 300$

Then $f_2(i) = \min_{x} \{c(x) + 1.5(i + x-300) + f_2(i + x-300)\}$

where
$$c(0) = 0$$
$$c(100) = 250 + 10(100) = 1,250$$
$$c(200) = 250 + 10(200) = 2,250$$
$$c(300) = 250 + 10(300) = 3,250$$
$$c(400) = 250 + 10(400) = 4,250$$
$$c(500) = 250 + 10(500) = 5,250$$
$$c(600) = 250 + 10(600) = 6,250$$
$$c(700) = 250 + 10(700) = 7,250$$
$$c(800) = 250 + 10(800) = 8,250$$

and $x \geq 300-i$. Note that during month 2 entering inventory cannot exceed $800 - 200 = 600$. Also note that during month 2 it would be foolish to produce more than $(300 + 300) - i = 600 - i$ units, because we would then have some inventory at the end of month 3.

Using these simplifications the necessary $f_2( )$ computations are as follows:

$$f_2(0) = \min \begin{cases} 3,250 + 0 + f_3(0) = 7100 \\ 4,250 + 150 + f_3(100) = 7,050 \\ 5,250 + 300 + f_3(200) = 7,000 \\ 6,250 + 450 + f_3(300) = 6,700* \end{cases}$$

$$x_2(0) = 600$$

$$f_2(100) = \min \begin{cases} 2,250 + 0 + f_3(0) = 6,100 \\ 3,250 + 150 + f_3(100) = 6,050 \\ 4,250 + 300 + f_3(200) = 6,000 \\ 5,250 + 450 + f_3(300) = 5,700* \\ x_2(100) = 500 \end{cases}$$

$$f_2(200) = \min \begin{cases} 1,250 + 0 + f_3(0) = 5,100 \\ 2,250 + 150 + f_3(100) = 5,050 \\ 3,250 + 300 + f_3(200) = 5,000 \\ 4,250 + 450 + f_3(300) = 4,700* \\ x_2(200) = 400 \end{cases}$$

131

$$f_2(300) = \min \begin{cases} 0 + 0 + f_3(0) = 3,850 \\ 1,250 + 150 + f_3(100) = 4,050 \\ 2,250 + 300 + f_3(200) = 4,000 \\ 3,250 + 450 + f_3(300) = 3,700* \\ \quad\quad x_2(300) = 300 \end{cases}$$

$$f_2(400) = \min \begin{cases} 0 + 150 + f_3(100) = 2,800 \\ 1,250 + 300 + f_3(200) = 3,000 \\ 2,250 + 450 + f_3(300) = 2,700* \end{cases}$$

$$x_2(400) = 200$$

$$f_2(500) = \min \begin{cases} 0 + 300 + f_3(200) = 1,750 \\ \\ 1,250 + 450 + f_3(300) = 1,700* \end{cases}$$

$$x_2(500) = 100$$

$$f_2(600) = 0 + 450 + f_3(300) = 450 \quad x_2(600) = 0$$

Now we compute $f_1(0)$ from

$$f_1(0) = \min_{x}\{c(x) + 1.5(i + x-200) + f_2(i + x-200)\}$$

where $x \geq 200$. Thus

$$f_1(0) = \min \begin{cases} 2,250 + 0 + f_2(0) = 8,950* \\ 3,250 + 150 + f_2(100) = 9,100 \\ 4,250 + 300 + f_2(200) = 9,250 \\ 5,250 + 450 + f_2(300) = 9,400 \\ 6,250 + 600 + f_2(400) = 9,550 \\ 7,250 + 750 + f_2(500) = 9,700 \\ 8,250 + 900 + f_2(600) = 9,600 \\ \quad\quad x_1(0) = 200 \end{cases}$$

Thus $f_1(0) = \$8,950$ and $x_1(0) = 200$ radios should be produced during month 1. This yields a month 2 inventory of $200 - 200 = 0$. Thus during month 2 we produce $x_2(0) = 600$ radios. At the beginning of month 3 the inventory will now be $0 + 600 - 300 = 300$. Hence during month 3 $x_3(300) = 0$ radios should be produced.

Note that the total production cost of this plan is 250 + 200(10) + 250 + 600(10) = 8,500 and Total Holding Cost = 1.5(300) = 450 (assessed on inventory at end of month 2). Thus total cost is 8500 + 450 = $8,950 = $f_1(0)$

SECTION 18.4

1. Let $f_t(i)$ be the maximum revenue attained from sites $t, t + 1, ..3$ if i million dollars can be invested in those sites. Let $x_t(i)$ be the amount of money (in millions) that should be assigned to site i in order to attain $f_t(i)$.

Then (all figures are in millions)

$f_3(4) = 15 \quad x_3(4) = 4$
$f_3(3) = 13 \quad x_3(3) = 3$
$f_3(2) = 8 \quad x_3(2) = 2$
$f_3(1) = 7 \quad x_3(1) = 1$
$f_3(0) = 3 \quad x_3(0) = 0$
$f_2(0) = 3 + f_3(0) = 6^* \quad x_2(0) = 0$

$$f_2(1) = \max \begin{cases} 3 + f_3(1) = 10^* & x_2(0) = 0 \\ 6 + f_3(0) = 9 \end{cases}$$

$$f_2(2) = \max \begin{cases} 3 + f_3(2) = 11 \\ 6 + f_3(1) = 13^* & x_2(2) = 1 \text{ or } 2 \\ 10 + f_3(0) = 13^* \end{cases}$$

$$f_2(3) = \max \begin{cases} 3 + f_3(3) = 16 \\ 6 + f_3(2) = 14 \\ 10 + f_3(1) = 17^* & x_2(3) = 2 \\ 12 + f_3(0) = 15 \end{cases}$$

$$f_2(4) = \max \begin{cases} 3 + f_3(4) = 18 \\ 6 + f_3(3) = 19^* \\ 10 + f_3(2) = 18 \\ 12 + f_3(1) = 19^* & x_2(4) = 1 \text{ or } 3 \\ 14 + f_3(0) = 17 \end{cases}$$

$$f_1(4) = \max \begin{cases} 4 + f_2(4) = 23 \\ 7 + f_2(3) = 24* \quad x_1(4) = 1 \\ 8 + f_2(2) = 21 \\ 9 + f_2(1) = 19 \\ 11 + f_2(0) = 17 \end{cases}$$

Thus $x_1(4) = 1$ million dollars should be allocated to site 1, $x_2(4 - 1) = 2$ million dollars should be allocated to site 2 and $x_3(3 - 2) = 1$ million dollars should be allocated to site 3.

2. We use (8). Let $g(w)$ be the maximum benefit that can be attained from a w pound knapsack. Also define $x(w)$ to be the type of item that attains the minimum in the recursion for a w pound knapsack.
Then (Type 0 item means nothing can fit in knapsack)
$g(0) = 0 \quad x(0) = 0$
$g(1) = 0 \quad x(1) = 0$
$g(2) = 2 \quad x(2) = 3$

$$g(3) = \max \begin{cases} 4 + g(0) = 4* \quad \text{(Put in Type 2 Item)} \\ \\ 2 + g(1) = 2 \quad \text{(Put in Type 3 Item)} \end{cases}$$

$x(3) = 2$

$$g(4) = \max \begin{cases} 5 + g(0) = 5* \\ 4 + g(1) = 4 \\ 2 + g(2) = 4 \end{cases}$$

$x(4) = 1$

$$g(5) = \max \begin{cases} 5 + g(1) = 5 \\ 4 + g(2) = 6* \\ 2 + g(3) = 6* \end{cases}$$

$x(5) = 2$

$$g(6) = \max \begin{cases} 5 + g(2) = 7 \\ 4 + g(3) = 8* \\ 2 + g(4) = 7 \end{cases}$$

x(6) = 2

$$g(7) = \max \begin{cases} 5 + g(3) = 9* \\ 4 + g(4) = 9 \\ 2 + g(5) = 8* \end{cases}$$

x(7) = 1 or 3

$$g(8) = \max \begin{cases} 5 + g(4) = 10* \\ 4 + g(5) = 10* \\ 2 + g(6) = 10* \end{cases}$$

x(8) = 2

We choose to begin by putting a Type 2 item in the knapsack. This leaves us with an 8 - 3 = 5 pound knapsack so we put an x(5) = Type 2 item in the knapsack. This leaves us with a 5 - 3 = 2 pound knapsack so we next put in an x(2) = Type 3 item to fill the knapsack. Thus the knapsack should be filled with two Type 2 items and one Type 3 item. There are, of course, other optimal solutions such as using two Type 1 items. Observe that we really did not need g(7) to determine g(8)!

SECTION 18.5

1. $f_5(1) = -800$, $f_5(2) = -600$, $f_5(3) = -500$
$f_4(3) = 560 + f_5(1) = -240*$ (Replace)

$$f_4(2) = \min \begin{cases} 460 + f_5(1) = -340 \text{ (Replace)} \\ 120 + f_5(3) = -380* \text{ (Keep)} \end{cases}$$

$$f_4(1) = \min \begin{cases} 260 + f_5(1) = -540* \text{ (Replace)} \\ 80 + f_5(2) = -520 \text{ (Keep)} \end{cases}$$

135

$f_3(3) = 560 + f_4(1) = 20*$ (Replace)

$$f_3(2) = \min \begin{cases} 460 + f_4(1) = -80 & \text{(Replace)} \\ 120 + f_4(3) = -120* & \text{(Keep)} \end{cases}$$

$$f_3(1) = \min \begin{cases} 260 + f_4(1) = -280 & \text{(Replace)} \\ 80 + f_4(2) = -300* & \text{(Keep)} \end{cases}$$

$f_2(3)$ cannot occur

$$f_2(2) = \min \begin{cases} 460 + f_3(1) = 160 & \text{(Replace)} \\ 120 + f_3(3) = 140* & \text{(Keep)} \end{cases}$$

$$f_2(1) = \min \begin{cases} 260 + f_3(1) = -40* & \text{(Replace)} \\ 80 + f_3(2) = -40* & \text{(Keep)} \end{cases}$$

$f_1(2)$ and $f_1(3)$ cannot occur.

$$f_1(1) = \min \begin{cases} 260 + f_2(1) = 220* & \text{(Replace)} \\ 80 + f_2(2) = 220* & \text{(Keep)} \end{cases}$$

$f_0(0) = 1,000 + 60 + f_1(1) = 1,280*$ (Keep)

To determine an optimal replacement policy note that at Time 1 we need to attain $f_1(1)$, so we may either keep or replace the machine. Suppose we choose to keep the machine. Then at Time 2 we need to attain $f_2(2)$, so we keep the machine. Then we need to attain $f_3(3)$, which requires that we replace the machine. Then to attain $f_4(1)$ we again replace the machine. Finally we replace the machine at Time 5.

SECTION 18.6

1. Let $f_t(i)$ = the maximum expected net profit
136

earned during years $t, t + 1, \ldots 10$ given that Sunco has i barrels of reserves at the beginning of year t. Then

$$f_{10}(i) = \max_{x} \{xp_{10} - c(x)\}$$

where x must satisfy $0 \leq x \leq i$.

For $t \leq 9$

$$(1) \quad f_t(i) = \max_{x} \{xp_t - c(x) + f_{t+1}(i + b_t - x)\}$$

where $0 \leq x \leq i$.

We use (1) to work backwards until $f_1(i_0)$ is determined.
(i_0 = number of barrels available at beginning of period 1). If discounting is allowed, let £ = the discount factor. Then we redefine $f_t(i)$ to be measured in terms of year t dollars. Then we replace (1) by (1')

$$(1') \quad f_t(i) = \max_{x} \{xp_t - c(x) + \beta f_{t+1}(i + b_t - x)\}.$$

where $0 \leq x \leq b$.

2a. Let d be the amount of money consumed during a year. If $u(d) = d^2$, then $u''(d) = 2 > 0$. This means that as consumption is increased, each additional dollar of consumption adds more to the person's utility. Most people do not behave this way. See Chapter 13 of OR. On the other hand if $u(d) = \ln d$, then $u''(d) = -1/d^2$, so each additional dollar of consumption adds less and less to Juli's utility. This is more consistent with the behavior of most people. In short, the behavior of few people can be described by convex utility functions.

2b. Let $f_t(d)$ be the maximum utility that can be earned during years $t, t + 1, 10$ given that d dollars are available at the beginning of year t (including year t income). During year 10 it makes sense to consume all available money (after all there is no future. Thus

$$f_{10}(d) = \ln d$$

For $t \leq 9$

$$f_t(d) = \max_c \{\ln c + f_{t+1}(1.1[d - c] + i)\}$$

where $0 \le c \le d$.

   We work backwards from the $f_{10}(\ )$'s to $f_1(D)$.

3. Let $f_t(i)$ be the minimum cost (in dollars) incurred during minutes t, $t + 1, \ldots 60$ given that i customers are present at the beginning of minute t (excluding minute t arrivals).
   Then $f_{60}(i) = \min_s \{c(s) + .10(i + x_{60} - c(s, i + x_{60})\}$

   For $t \le 59$

$$f_t(i) = \min_s \{c(s) + .10(i + x_t - c(s, i + x_t))$$
$$+ f_{t+1}(i + x_t - c(s, i + x_t))\}$$

where s is a member of $\{0,1,2,\ldots\}$.

   We begin by computing the $f_{60}(\ )$ and work backwards until
$f_1(0)$ has been computed.

Section 18.7

1. If initial inventory is 200 units then, in effect, we may assume that our initial inventory of 200 units is used to meet period 1 demand and assume that period 1 has a demand of $220 - 200 = 20$ units. Ignoring the \$2 per unit cost (which is independent of the production schedule) we obtain

$f_6 = 0$
$f_5 = 250*$ (produce for period 5)

$f_4 = \min \begin{cases} 250 + f_5 = 500* & \text{(produce for period 4)} \\ 250 + 270 + f_6 = 520 & \text{(produce for periods 4 and 5)} \end{cases}$

$f_3 = \min \begin{cases} 250 + f_4 = 750 \\ 250 + 140 + f_5 = 640* \\ 250 + 140 + 2(270) + f_6 = 930 \end{cases}$

$$f_2 = \min \begin{cases} 250 + f_3 = 890* \\ 250 + 360 + f_4 = 1110 \\ 250 + 360 + 2(140) + f_5 = 1140 \\ 250 + 360 + 2(140) + 3(270) + f_6 = 1700 \end{cases}$$

$$f_1 = \min \begin{cases} 250 + f_2 = 1140* \\ 250 + 280 + f_3 = 1170 \\ 250 + 280 + 2(360) + f_4 = 1750 \\ 250 + 280 + 2(360) + 3(140) + f_5 = 1920 \\ 250 + 280 + 2(360) + 3(140) + 4(270) + f_6 \end{cases}$$

$=2750$

Thus produce 20 units during period 1, $d_2 = 280$ units during period 2, $d_3 + d_4 = 500$ units during period 3, and $d_5 = 270$ units during period 5.

If initial inventory is 400 units, then the computations of f_2- f_6 remain unchanged. In computing f_1, however, we do not have the option to just produce for period 1. Thus

$$f_1 = \min \begin{cases} 250 + 280 + f_3 = 1170* \\ 250 + 280 + 2(360) + f_4 = 1750 \\ 250 + 280 + 2(360) + 3(140) + f_5 = 1920 \\ 250 + 280 + 2(360) + 3(140) + 4(270) + f_6 = \end{cases}$$

2750

Thus during period 1 produce enough to meet demands for periods 1 and 2 $(220 + 280 - 400 = 100 \text{ units})$. In period 3 produce $360 + 140 = 500$ units. In period 5 produce 270 units.

Section 18.8

1. From EXCEL(file S18_8_1.xls) we find $f_1(0) = 8950$. From cell E17 we should produce 200 radios in month 1. Then we seek $f_2(0+200-200) = 6700$. From cell I16 we find that 600 radios should be produced during month 2. Then we seek $f_3(0+600-300)$

= 0 and 0 radios should be produced during month 3.

Chapter 19 Solutions

SECTION 19.1

2. Let $f_t(d)$ be the maximum expected NPV earned from investments $t, \ldots 3$ given that d million dollars are available for investment in investments $t, \ldots 3$. Also define $x_t(d)$ = Amount of money (in millions) that should be assigned to investment t in order to attain $f_t(d)$. Finally, define $r_t(x)$ to be the expected NPV(in millions) obtained from investment t if x million dollars are invested in Investment t.

Then $f_3(d) = r_3(d) \quad x_3(d) = d$

and for $t \leq 2$

$$f_t(d) = \max_{x} \{r_t(x) + f_{t+1}(d-x)\}$$

where x is a member of $\{0, 1, \ldots d\}$.

We begin by computing the $r_t(x)$.

$r_1(0) = 0$
$r_1(1) = 2(.6) + 4(.3) + 5(.1) = 2.9$
$r_1(2) = 4(.5) + 6(.3) + 8(.2) = 5.4$
$r_1(3) = 6(.4) + 7(.5) + 10(.1) = 6.9$
$r_1(4) = 7(.2) + 9(.4) + 10(.4) = 9$
$r_2(0) = 0$

$r_2(1) = 1(.5) + 2(.4) + 4(.1) = 1.7$
$r_2(2) = 3(.4) + 5(.4) + 6(.2) = 4.4$
$r_2(3) = 4(.3) + 6(.3) + 8(.4) = 6.2$
$r_2(4) = 3(.4) + 8(.3) + 9(.3) = 6.3$
$r_3(0) = 0$
$r_3(1) = .2(0) + .6(4) + .2(5) = 3.4$
$r_3(2) = 4(.4) + 6(.4) + 7(.2) = 5.4$
$r_3(3) = 5(.3) + 7(.4) + 8(.3) = 6.7$
$r_3(4) = 6(.1) + 8(.5) + 9(.4) = 8.2$
Then

$f_3(4) = 8.2 \quad x_3(4) = 4$
$f_3(3) = 6.7 \quad x_3(3) = 3$
$f_3(2) = 5.4 \quad x_3(2) = 2$
$f_3(1) = 3.4 \quad x_3(1) = 1$

$f_3(0) = 0 \quad x_3(0) = 0$

$$f_2(4) = \max \begin{cases} 0 + f_3(4) = 8.2 \\ 1.7 + f_3(3) = 8.4 \\ 4.4 + f_3(2) = 9.8* \quad x_2(4) = 2 \\ 6.2 + f_3(1) = 9.6 \\ 6.3 + f_3(0) = 6.3 \end{cases}$$

$$f_2(3) = \max \begin{cases} 0 + f_3(3) = 6.7 \\ 1.7 + f_3(2) = 7.1 \\ 4.4 + f_3(1) = 7.8* \quad x_2(3) = 2 \\ 6.2 + f_3(0) = 6.2 \end{cases}$$

$$f_2(2) = \max \begin{cases} 0 + f_3(2) = 5.4* \quad x_2(2) = 0 \\ 1.7 + f_3(1) = 5.1 \\ 4.4 + f_3(0) = 4.4 \end{cases}$$

$$f_2(1) = \max \begin{cases} 0 + f_3(1) = 3.4* \quad x_2(1) = 0 \\ 1.7 + f_3(0) = 1.7 \end{cases}$$

$f_2(0) = 0 + f_3(0) = 0 \quad x_2(0) = 0$

$$f_1(4) = \max \begin{cases} 0 + f_2(4) = 9.8 \\ 2.9 + f_2(3) = 10.7 \\ 5.4 + f_2(2) = 10.8* \\ 6.9 + f_2(1) = 10.3 \quad x_1(4) = 2 \\ 9 + f_2(0) = 9 \end{cases}$$

Assign $x_1(4) = 2$ million to Investment $1, x_2(4 - 2) = 0$ million dollars to Investment 2 and $x_3(2 - 0) = 2$ million dollars to Investment 3. A total expected NPV of 10.8 million dollars will be obtained.

SECTION 19.2

2. Now $f_3(3) = 0$, $x_1(3) = 0$, $f_3(2) = 0$, $x_2(2) = 0$, $f_3(1) = 5$, $x_3(1) = 1$ $f_3(0) = 7$, $x_3(0) = 2$

$$f_2(3) = \min \begin{cases} 0 + 3 + 1/2\{f_3(2) + f_3(1)\} = 5.5* \\ 5 + 5 + 1/2\{f_3(3) + f_3(2)\} = 10 \end{cases}$$

$$f_2(2) = \min \begin{cases} 0 + 1 + 1/2\{f_3(0) + f_3(1)\} = 7* \\ 5 + 3 + 1/2\{f_3(1) + f_3(2)\} = 10.5 \\ 7 + 5 + 1/2\{f_3(2) + f_3(3)\} = 12 \end{cases}$$

$$f_2(1) = \min \begin{cases} 5 + 1 + 1/2\{f_3(1) + f_3(0)\} = 12* \\ 7 + 3 + 1/2\{f_3(2) + f_3(1)\} = 12.5 \\ 9 + 5 + 1/2\{f_3(3) + f_3(2)\} = 14 \end{cases}$$

$$f_2(0) = \min \begin{cases} 7 + 1 + 1/2\{f_3(1) + f_3(0)\} = 14* \\ 9 + 3 + 1/2\{f_3(2) + f_3(1)\} = 14.5 \\ 11 + 5 + 1/2\{f_3(3) + f_3(2)\} = 16 \end{cases}$$

$$f_1(1) = \min \begin{cases} 5 + 1 + 1/2\{f_2(1) + f_2(0)\} = 19 \\ 7 + 3 + 1/2\{f_2(2) + f_2(1)\} = 19.5 \\ 9 + 5 + 1/2\{f_2(3) + f_3(2)\} = 20.25 \end{cases}$$

Thus $x_1(1) = 1$, $x_2(1) = 1$, $x_2(0) = 2$, $x_3(1) = 1$, $x_3(0) = 2$. After observing each period's demand, we may determine the optimal production level for the period. Minimum expected cost is \$19.

SECTION 19.3

1. Let f_0 be the probability that Ulanowsky will win the match if he plays optimally. Let $f_1(W)$ be the probability that Ulanowsky will win the match given that he has won the first game (similarly define $f_1(L)$ and $f_1(D)$). Let L = Loss and D = Draw.
 Then define $f_2(WL)$ to be the probability that Ulanowsky wins the match given that he has won the first game and lost the second game. Similarly define $f_2(LW)$, $f_2(DL)$, etc.
Then
$f_2(WW) = f_2(WD) = f_2(DW) = 1$
$f_2(LL) = f_2(LD) = f_2(DL) = 0$
$f_2(DD) = f_2(WL) = f_2(LW) = .45$. This follows because if the match is tied after two games then Ulanowsky must play boldly or he will have no chance of winning the tie-breaking game.
Working backwards we find that

$$f_1(W) = \max \begin{cases} .45f_2(WW) + .55f_2(WL) = .6975 \\ .10f_2(WL) + .90f_2(WD) = .945* \end{cases}$$

$$f_1(L) = \max \begin{cases} .45f_2(LW) + .55f_2(LL) = .2025* \\ .10f_2(LL) + .90f_2(LD) = 0 \end{cases}$$

$$f_1(D) = \max \begin{cases} .45f_2(DW) + .55f_2(DL) = \\ .10f_2(DL) + .90f_2(DD) = .405 \end{cases}$$

Now

$$f_0 = \max \begin{cases} .45f_1(W) + .55f_1(L) = .537* \\ .10f_1(L) + .90f_1(D) = .425 \end{cases}$$

Thus Ulanowsky should play boldly during the first game. If he wins the first game, then he should play conservatively in the second game. If he loses the first game, then play boldly during the second game. If there is a tiebreaking game, he should play boldly. Ulanowsky's chance of winning the match is $f_0 = .537$.

The reason why Ulanowsky is more likely to win the match than Smithson is that he can adjust his strategy after observing the outcome of the first game; Smithson has no such option.

SECTION 19.4

1. Let $f_t(w)$ be the minimum expected shortage costs due to type $t, t + 1, \ldots T$ fuel cells if w fuel cells can be taken on the flight.
Then

$$f_T(w) = c_T \sum_{z>w} p_T(z)$$

and for $t \leq T-1$

$$f_t(w) = \min_{x_t} \left\{ c_t \sum_{z > x_t} p_t(z) + f_{t+1}(w - x_t) \right\}$$

where x_t is a member of $\{0, 1, 2, \ldots w\}$.
Work backwards until $f_1(W)$ has been determined.

2. Let $f_t(d)$ be the maximum expected asset position of the firm at the end of year 10 given that at the beginning of year t the firm has d dollars in assets.

Then

$$f_{10}(d) = \max_I \left\{ p \sum_y q_y(d + i + y) + (1-p) \sum_y q_y(d - i + y) \right\}$$

$$f_t(d) = \max_i \left\{ p \sum_y q_y f_{t+1}(d + i + y) + (1-p) \sum_y q_y f_{t+1}(d - i + y) \right\}$$

We work backwards until $f_1(10,000)$ has been computed.

3. Let $f_t(i)$ be the minimum expected discounted cost (in month t dollars) incurred in operating the machine during months $t, t + 1, \ldots T$ given that at the beginning of month t we have a state i machine.
   Note that \$1 now = \$1.01 a month from now so \$.99 now = \$1/1.01 now = \$1 a month from now

Then
$$f_T(i) = \min \begin{cases} R + c(0) \\ c(i) \end{cases}$$

and for $t \leq T-1$

$$f_t(t) = \min(R + c(0) + .99 \sum_{j=0}^{j=\infty} p_{oj} f_{t+1}(j), c(i) + .99 \sum_{j=0}^{j=\infty} p_{ij} f_{t+1}(j)).$$

First term corresponds to replacing and second term to not replacing.

We work backwards until we compute $f_1(i_0)$.

Chapter 20 Solutions

1. After the first service completion, I enter service. By the no memory property of the exponential distribution, I am no more or less likely to complete service before any of the remaining customers. Hence there is a 1/7 chance that I will be the last customer to complete service, and a $1 - 1/7 = 6/7$ chance that I will not be the last customer to complete service.

2. A typical segment of time looks like
   30 minute interval
   60 minute interval
   120 minute interval
   120 minute interval

   Thus the chance of arriving during segments of various lengths are
   $30/330 = 1/11$ for a 30 minute segment
   $60/330 = 2/11$ for a 1 hour segment
   $240/330 = 8/11$ for a 2 hour segment
   Hence the expected waiting time for next bus is given by $1/11[.5(30)] + 2/11[.5(60)] + 8/11[.5(120)] = 555/11 = 50.45$ minutes
   Note: Average Time between Buses $= 1/4(30) + 1/4(60) + 1/2(120) = 82.5$ minutes and $(1/2)82.5 \neq 50.45$

5. The probability that a person has no accident during a year is $POISSON(0, 0.022, 1) = .978$. Thus in a sense during a given year 97.8% of all drivers perform better than average.

SECTION 20.3

2a.   Let the state be the number of working bulbs.
Possible states are 0, 1, and 2.   Birth = a bulb
is repaired while Death = bulb burns out.   Then
the birth-death parameters are as follows:

$\lambda_0 = 1/2 + 1/2 = 1$ $\mu_0 = 0$
$\lambda_1 = 1/2$ $\mu_1 = 1/22$
$\lambda_2 = 0$ $\mu_2 = 1/22 + 1/22 = 1/11$

Steady state probabilities may be found from
$\pi_0 = \pi_1/22$, $\pi_1/2 + \pi_1/22 = \pi_0 + \pi_2/11$, $\pi_2/11 = \pi_1/2$,
$\pi_0 + \pi_1 + \pi_2 = 1$
or $\pi_1 = 22\pi_0$, $\pi_2 = 121\pi_0$.   Thus $\pi_0(1 + 22 + 121) = 1$
or
$\pi_0 = 1/144$, $\pi_1 = 11/72$, $\pi_2 = 121/144$.

2b. $\pi_2 = 121/144$

2c. $\pi_0 = 1/144$

3a. From file S20_3_3.xls in cell G36 we find that
the expected number of pizza restaurants is 20.47.

3b. From file s20_3_3.xls we find that (cell H24)
there will be more than 20 pizza restaurants 99.9%
of the time.

SECTION 20.4

1.   We have an M/M/1 system with
$\lambda = 10$ customers/minute $\mu = 12$ customers/minute,
and $\rho = 10/12 = 5/6$.

a. $1 - \pi_0 = 5/6$

b. $L_q = \dfrac{\rho^2}{1-\rho} = \dfrac{25/36}{1/6} = 25/6$ customers.

c. $L = \dfrac{\rho}{1-\rho} = \dfrac{5/6}{1/6}$ 5 customers

$W = L/\lambda = 5/10 = 1/2$ minute $= 1/120$hr.

2. For the slow copier service cost/hour = \$4, $\lambda = 4$
customers/hour, $\mu = 6$ customers/hour

$W = \dfrac{1}{\mu - \lambda} = 1/2$ hour

delay cost/hour = (4 customer/hr.) (1/2hr.)
(\$15/cust.- hr.) = \$30/hr.
total cost/hour = \$34.
For fast copier, service cost/hour = \$15/hour, $\lambda = 4$ cust./hr., and $\mu = 10$ cust/hr.

$W = \dfrac{1}{\mu - \lambda} = 1/6$ hr.

delay cost/hr. = (4 cust./hr.) (1/6 hr.)
(\$15/cust.-hr.)= \$10/cust.-hr.
total cost/hr. = \$25/hr. Thus the fast copier is
better.

3. Let λ, μ, L, W and π_j refer to the original
system and 2λ, 2μ, L', W' and π' refer to the new
system. Since the new system has the same ρ as
the old system $\pi_j = \pi_j'$, and the steady state
probabilities remain unchanged. Since $L = \rho/(1-\rho)$, we will have $L = L'$ and expected queue length
is unchanged. Finally, since $W = L/\lambda$ we have that
$W' = L'/2\lambda = W/2$. Thus expected waiting time for
new system is half expected waiting time for old
system.

SECTION 20.5

1. We have an M/M/1/3 system with $\lambda = 3$
customers/hour and $\mu = 2$ customers/hour.

      a. We seek $\lambda(1-\pi_3)$. Since

$$\pi_0 = \frac{1 - 3/2}{1 - (3/2)^4} = \frac{16}{130}$$

$\pi_3 = (3/2)^3 16/130) = .415$. Thus
$\lambda(1 - \pi_3) = 1.75$ customers/hour

b. $1 - \pi_0 = 114/130$

2. We have an M/M/1/4 system with $\mu = 15$ cars/hour and $\lambda = 40$ cars/hour.

2a. $L_q = L - L_s$ $\rho = 40/15 = 2.67$

$$L = \frac{2.67 \ [1 - 5(2.67)^4 + 4(2.67)^5]}{(1 - 2.67^5) \ (1 - 2.67)} = 3.44$$

$$\pi_0 = \frac{1 - 2.67}{1 - 2.67^5} = .012 \quad L_s = 1 - \pi_0 = .99$$

$L_q = 3.44 - .99 = 2.45$

2b. We seek $\lambda(1 - \pi_4) = 40(1 - .61) = 15.6$ cars/hour
$(\pi_4 = (2.67)^4(.012) = .61)$

2c. We seek $W = \dfrac{L}{\lambda(1 - \pi_4)} = \dfrac{3.44}{40(1 - .61)} = .22$ hours

SECTION 20.6

1. $\lambda = 18$ customers/hour, $\mu = 15$ customers/hour. If we have s servers
$\rho = 18/15s$, so at least 2 servers are needed to ensure that $\rho < 1$.

For s = 2 ρ = .60 and W = $\dfrac{.45}{30-18}$ + $\dfrac{1}{15}$ = .104 hours and

DelayCostPer Hour=($15 per cust-hr.(18 cust/hr.)(.104 hr.) = $28.08.

Total Cost/Hour = 40 + 28.08 = $68.08
For 3 servers ρ = 18/45 = .40 and

W = $\dfrac{.14}{45-18}$ + $\dfrac{1}{15}$ = .072 hours

$\dfrac{\text{Delay Cost}}{\text{cust.-hr.}}$ = $\dfrac{\$15}{\text{hr.}}$ $\dfrac{18 \text{ cust.}}{\text{hr.}}$ (.072 hours)

=$19.44.
Thus for s = 3 Total Hourly Cost = 60 + 19.44 = $79.44
Two servers is better than three servers. Four or more servers incurs a service cost of $80 or more, so 2 servers is optimal.

2. λ = 50 customers/day, μ = 60 customers/day.
For s servers Expected Cost/Day = 100's + λ(100W).
For s = 1 W = 1/(60 - 50) = 1/10 days and
Expected Cost/Day = 100 + 50 (100) (1/10) = $600
For s = 2 ρ = 50/120 = .41 Then interpolation yields P(j≥2) = .24

W = $\dfrac{.24}{120-50}$ + $\dfrac{1}{60}$ = .020 days

and Expected Cost/Day = 200 + 50(100) (.020) = $300/day
Since three or more servers must cost at least $300/day, two servers must be optimal.

3. Finance is an M/M/1 system with λ = 20 letters/day and μ = 25 letters/day. Thus $W_{Finance}$ = 1/(25-20) = 1/5 day.

151

Marketing is an M/M/1 system with λ = 15 letters/day and μ = 25 letters/day. Thus $W_{Marketing}$ = 1/(25-15) = 1/10 day.
The pooled system is an M/M/2 system with λ = 20 + 15 = 35 letters/day and μ = 25 letters/day. Then ρ = 35/2(25) = .70 and P(j$\geq$2) = .57. Then L_q = (.57).70/(1 - .70) = 1.33 letters and

$$W = \frac{L_q}{\lambda} + \frac{1}{\mu} = \frac{1.33}{35} + \frac{1}{25} = .078 \text{ days}$$

Thus both finance and marketing are better off under the pooled system. The reason for this is that if the two departments pool their servers each typist will spend much less time waiting for work and the system will be more efficient. To answer part (d) we note that we have an M/M/2 system with λ = 35 letters/day and μ = 25 letters/day. Then ρ = 35/50 = .70 and P(j$\geq$2) = .57. Then

$$P(W>.20) = e^{-25(.20)}[1 + (.57)\frac{1-e^{-25(.20)(-.4)}}{-.4}]$$

= .068.

SECTION 20.7

1. λ = 100 members/week W = 52 weeks. Then L = λW = 5200 members.

SECTION 20.8

1. Use Pollachek-Khinchine Formulas with σ^2 = 0, λ = 20 customers/hr. and μ = 30 customers/hour. Then

$$L_q = \frac{(20)^2(0) + (2/3)^2}{2(1 - 2/3)} = 2/3 \text{ customers and}$$

$W_q = L_q/\lambda$ = (2/3)/20 = 1/30 hour = 2 minutes.

1. We are comparing two machine repair systems
System 1: λ = 1/5 machine/day μ = .4 machines/day
K = 5, R = 3, ρ = .5
System 2: λ = 1/5 machine/day μ = 1.2 machines/day
K = 5 R = 1 ρ = 1/6
Let L = number of machines in bad condition.
Let's compare L for both systems. For System 1

$$\pi_1 = \binom{(5)}{(1)} (1/2) \ \pi_0 = (5/2) \pi_0$$

$$\pi_2 = \binom{(5)}{(2)} (1/2)^2 \pi_0 = (5/2) \pi_0$$

$$\pi_3 = \binom{(5)}{(3)} (1/2)^3 \pi_0 = (5/4) \pi_0$$

$$\pi_4 = \binom{(5)}{(4)} (1/2)^4 \ \frac{4!}{3!(3)} \ \pi_0 = (5/12) \pi_0$$

$$\pi_5 = \binom{(5)}{(5)} 1/2)^5 \ \frac{5!}{3!3^2} \pi_0 = (5.72) \pi_0$$

Then $\pi_0 [1 + (5/2) + (5/2) + (5/4) + (5/12) + (5/72]$
= 1 or
π_0 = 72/557, π_1 = 180/557, π_2 = 180/557, π_3 = 90/557,
π_4 = 30/557, π_5 = 5/557

$$\text{and} \ L = \frac{0(72) + 1(180) + 2(180) + 3(90) + 4(30) \ 5(5)}{557}$$

$$= 1.71 \text{ machines}$$

153

For System 2

$$\pi_1 = \binom{(5)}{(1)} (1/6)\,\pi_0 = (5/6)\,\pi_0$$

$$\pi_2 = \binom{(5)}{(2)} (1/6)^2 2!\,\pi_0 = (20/36)\,\pi_0$$

$$\pi_3 = \binom{(5)}{(3)} (1/6)^3 3!\,\pi_0 = (60/216)\,\pi_0$$

$$\pi_4 \binom{(5)}{(4)} (1/6)^4 4!\,\pi_0 = (5/54)\,\pi_0$$

$$\pi_5 = \binom{(5)}{(5)} (1/6)^5 5!\,\pi_0 = (20/1296)\,\pi_0 \quad \text{Thus}$$

$\pi_0(1 + (5/6) + (20/36) + (60/216) + (5/54) + (20/1296) = 1$ or
$\pi_0 = 1296/3596, \pi_1 = 1080/3596, \pi_2 = 720/3596, \pi_3 = 360/3596,$
$\pi_4 = 120/3596$ and $\pi_5 = 20/3596$. Then for System 2

$$L = \frac{1(1080)+2(720)+3(360)+4(120)+5(20)}{3596}$$

$$= 1.16 \text{ machines}$$

Hence System 2 is superior to System 1.

SECTION 20.10

1. Option 1 is an M/M/2 system with $\lambda = 4.8$ applicants/hr. and $\mu = 4$ applicants/hour. Option 2 is the following series queuing system For Option 1 $\rho = 4.8/12 = .40$ and $P(j \geq 3) = .14$. Then

$$L_q = \frac{.14(.40)}{1 - .40} = 0.09 \text{ applicants}$$

$L + L_q + \lambda/\mu = 0.09 + 4.8/4 = 1.29$ applicants

$W = L/\lambda = 1.29/4.8 = .27$ hours
For Option 2 let W_1 be expected time in Stage 1 and W_2 be expected time in Stage 2. Since Stage 1 is an $M/M/\infty$ system, $W_1 = 1/\mu = 1.08$ hours.
To find W_2 note that Stage 2 is an $M/M/3$ with $\rho = 4.8/45 - .11$.
Then $P(j \geq 3) = 0$ so L_q for Stage 2 is 0. Then $L = L_q$ for Stage 2 is 0. Then $L = L_q + \lambda/\mu = 0 + 4.8/15 = .32$ customers and $W_2 = L/\lambda = .32/4.8 = .07$ hours. Thus $W_1 + W_2 = 1.08 + .07 = 1.15$ hours. Hence Option 1 results in a much smaller customer waiting time than Option 2.

6a. $\lambda_1 = 10 + .1\lambda_2$, $\lambda_2 = \lambda_1 + .2\lambda_3$, and $\lambda_3 = .9\lambda_2$. Solving these equations we find that $\lambda_1 = 11.39$, $\lambda_2 = 13.89$, and $\lambda_3 = 12.5$. Each server is busy a fraction $\rho_i = \lambda_i/20$ of the time. Thus
Server is busy $11/(39/20) = 57\%$ of the time
Server 2 is busy $13.89/20 = 69.5\%$ of the time
Server 3 is busy $12.5/20 = 62.5\%$ of the time.

6b. Expected number of jobs at station i $= \lambda_i/(\mu - \lambda_i)$. Thus
Expected Number of jobs at Station 1 $= 11.39/(20 - 11.39) = 1.32$
Expected number of jobs at Station 2 $= 13.89/(20 - 13.89) = 2.27$
Expected number of jobs at Station 3 $= 12.5/(20 - 12.5) = 1.67$
Total expected number of jobs in the system = 1.32 + 2.27 + 1.67 = 5.26 customers.

6c. $W = L/\lambda$, where $\lambda = 10$ jobs/hour. Thus $W = 5.26/10 = .526$ hours.

SECTION 20.11

1. We have a BCC problem so we use the graph with λ = 24 alarms/hr., μ = 3 alarms/hr. and λ/μ = 8 From the graph we see that to have at most a .01 chance of turning away an alarm requires at least 15 servers.

2. This is a BCC problem with λ = 480/8 = 60 calls/hr. and $1/\mu$ = .10 hours. Thus λ/μ = 6. Looking at the BCC graph we see that at least 13 servers are necessary. We need to assume that the time between phone calls (but not necessarily the length of a phone call) is exponentially distributed.

SECTION 20.12

1. Estimate μ by $\dfrac{24}{4 + 6 + 5 + \ldots + 130}$ = .0234
Use four categories each having an expected value of 6 observations.
P(Service Time$\geq$t) is estimated to be $1 - e^{-.0234t}$. To determine the boundaries of the categories solve
$1 - e^{-.0234t}$ = .25 or t = 12.3
$1 - e^{-.0234t}$ = .50 or t = 29.6
$1 - e^{-.0234t}$ = .75 or t = 59.2

	Range	Observed#	Expected #
Category 1	$0 \geq t < 12.3$	8	6
Category 2	$12.3 \geq t < 29.6$	4	6
Category 3	$29.6 \geq t < 59.2$	5	6
Category 4	$t > 59.2$	7	6

Chi-Square Observed =
$$\frac{(8-6)^2}{6} + \frac{(4-6)^2}{6} + \frac{(5-6)^2}{6} + \frac{(7-6)^2}{6}$$

= 1.67.
Since Chi-square$_{4-1-1}$(.05) = 5.99 we accept the hypothesis that the length of a customer's phone call follows an exponential distribution.

Section 20.13

1. See file S20_13_1.xls.

SECTION 20.15

1. Using (60) and expressing arrival and service rates on an hourly basis we obtain
ρ_1 = .25/1 = 1/4, ρ_2 = .0625/.25 = .25, ρ_3 = .625/2 = 5/16
a_0 = 0, a_1 = 1/4, a_2 = 1/2, a_3 = 13/16. From (60)

$$W_{q1} = \frac{(1/4)(2)/2 + (1/16)(20)/2 + (5/8)(1/2/2)}{(1-0)(1-1/4)} = 11/8 \text{ hours}$$

$$W_{q2} = \frac{(1/4)(2)/2 + (1/16)(20)/2 + (5/8)(1/2/2)}{(1-1/4)(1-1/2)} = 11/4 \text{ hours}$$

CHAPTER 21 Solutions

SECTION 21.4

1. Random Number Allocations for the Inter-arrival Time Distribution

Inter-arrival Time	Probability	Cumulative Distribution	Random Number Ranges
1 minute	0.20	0.20	00 - 19
2 minutes	0.30	0.50	20 - 49
3 minutes	0.35	0.85	50 - 84
4 minutes	0.15	1.00	85 - 99

Random Number Allocations for the Service Time Distribution

Service Time	Probability	Cumulative Distribution	Random Number Ranges
1 minute	0.35	0.35	00 - 34
2 minutes	0.40	0.75	35 - 74
3 minutes	0.25	1.00	75 - 99

If we use the first 2 columns in Table 5 (column 1 for the inter-arrival times and column 2 for the service times), we will generate the following inter-arrival and service times.

Customer Number	Inter-arrival Time	Service Time
1	-	2
2	4	3
3	1	2
4	2	3
5	2	3
6	3	2
7	4	2
8	2	1
9	2	1
10	4	3
11	3	3
12	2	1
13	2	3
14	2	2
15	1	1
16	1	1
17	3	1
18	3	2
19	4	2
20	3	1
21	2	1
22	3	1
23	1	1
24	2	3
25	2	1

If we now use these inter-arrival and service times to simulate the single server queuing model for the first 25 customers, we get a total of 19 minutes in waiting time. Therefore, the expected waiting time will be 0.76 minutes per customer. In other words, each customer waits 0.76 minutes on the average. This, however, is not likely to be a very accurate estimate of the expected waiting time because of the influence of the random numbers. We would have to simulate a lot more customers to get a more accurate estimate of the expected waiting time. Also note that your answer may vary according to the random numbers used in your computations.

SECTION 21.5

1. This function may be represented graphically as follows:

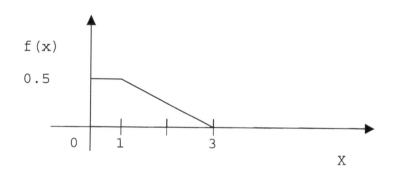

(a). In Step 1 of the ITM, we calculate the cdf. This cdf is represented formally by the function

$$F(x) = \begin{cases} 0 & x<0 \\ x/2 & 0 \leq x \leq 1 \\ -x^2/8 + 3x/4 - 1/8 & 1 \leq x \leq 3 \\ 1 & x \geq 3 \end{cases}$$

In Step 2 we generate a random number, r. In Step 3 we set $F(x) = r$ and solve for x. If r<.5, we use $F(x) = x/2$. Otherwise we use $F(x) = -x^2/8 + 3x/4 - 1/8$. If we now solve these equations, we get the following algorithm for the ITM:

161

```
    If (r≤.5) Then x= 2r
  Else x = 3 - (8 -8r)^{1/2}
  Endif
```

(b) For the ARM we solve the problem as follows

```
    Step 1: Set M = .5
    Step 2: Generate r₁ and r₂
    Step 3 x* = 3r₁
    Step 4: If (x*≤1) Then f(x*) = .5 Else f(x*) =
3/4 - 3r₁/4
            Endif
    Step 5:  If (x* ≤ 1) then
            Accept x* as the random variate
        Else
            If (r₂ ≤ 3/2 (1 - r₁)) Then
                Accept x* as the random variate
            Else
                Reject x* and go back to Step 2
            Endif
        Endif
```

The solution may be summarized by the following algorithm:

Generate 2 random numbers r_1 and r_2. Set $x^* = 3r_1$. If $(x^* \leq 1)$ Then accept x^* as the rnadom variate.

```
Else
    If (r₂≤3(1-r₁)/2. Then accept x* as the random
variate.
    Else
        Reject x* and repeat the process
        Endif
    Endif
```

2. This function may be represented graphically as follows:

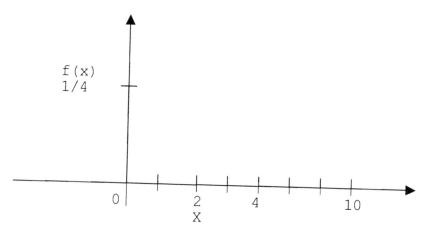

In Step 1 of the ITM, we first calculate the cdf for this function. This is represented formally as

$$F(x) = \begin{cases} 0 & x<2 \\ x^2/16 - x/4 + 1/4 & 2\le x \le 4 \\ -x^2/48 + 10x/24 - 52/48 & 4 \le x \le 10 \\ 1 & x\ge 10 \end{cases}$$

Next, we generate a random number, r. In Step 3 we set $F(x) = r$ and solve for x. For this function, Step 3 gives us the equations

$$x^2/16 - x/4 + 1/4 = r \qquad \text{for } 0 \le r \le 0.25 \tag{1}$$

and $-x^2/48 + 10x/24 - 52/48 = r$ for $0.25 < r \le 1$ (2)

Both these are quadratic equations and may be solved either by completing squares or by using the quadratic formula. We use the former. To solve equation (1), we multiply the equation by 16 and rearrange the terms to get

$$x^2 - 4x + 4 = 16r$$

This gives us

$$(x - 2)^2 = 16r$$

If we now take the square root of both sides, we get

$$x = 2 \pm \sqrt{16r}$$

as the solution. Since the random variable is

163

defined only for values of $x \geq 2$, only $x = 2 + 4 \sqrt{r}$ is feasible.

Using the same procedure for equation (2), we get

$x = 10 - 4 \sqrt{3-3r}$ as the solution. Thus, our algorithm for generating random variates from this distribution may be summarized as

Generate a random number, r

$\quad\quad\quad$ If $(r \leq 0.25)$ Then

$\quad\quad\quad\quad\quad x = 2 + 4 \sqrt{r}$

$\quad\quad\quad$ Else

$\quad\quad\quad\quad\quad x = 10 - 4 \sqrt{3 - 3r}$

$\quad\quad\quad$ Endif

SOLUTIONS TO CHAPTER 21 REVIEW PROBLEMS

1. The generator is

$\quad\quad\quad\quad X_{i+1} = (a\ X_i + c)$ module m

with $a = 17$, $c = 43$, $m = 100$, $X_o = 31$

Compute $X_1 = (17*31 + 43)$ mod 100

$\quad\quad\quad\quad = 70$

Deliver $R_1 = 0.70$

Compute $X_2 = (17*70 + 43)$ mod 100

$\quad\quad\quad\quad = 33$

Deliver $R_2 = 0.33$

Compute $X_3 = (17*33 + 43)$ mod 100

$\quad\quad\quad\quad = 4$

Deliver $R_3 = 0.04$

Compute $X_4 = (17*4 + 43)$ mod 100

$\quad\quad\quad\quad = 11$

Deliver $R_4 = 0.11$

Compute $X_5 = (17*11 + 43)$ mod 100

$\quad\quad\quad\quad = 30$

Deliver $R_5 = 0.30$

Compute $X_6 = (17*30 + 43)$ mod 100

$\quad\quad\quad\quad = 53$

Deliver $R_6 = 0.53$

Compute $X_7 = (17*53 + 43)$ mod 100

$\quad\quad\quad\quad = 44$

Deliver $R_7 = 0.44$

Compute $X_8 = (17*44 + 43)$ mod 100

$\quad\quad\quad\quad = 91$

```
Deliver R₈  = 0.91
Compute X₉  = (17*91 + 43) mod 100
           = 90
Deliver R₉  = 0.90
Compute X₁₀ = (17*90 + 43) mod 100
           = 73
Deliver R₁₀ = 0.73
```

Deliver R_8 = 0.91
Compute X_9 = (17*91 + 43) mod 100
       = 90
Deliver R_9 = 0.90
Compute X_{10} = (17*90 + 43) mod 100
       = 73
Deliver R_{10} = 0.73

Chapter 22 Solutions

See .igx files for Process model diagrams for each problem.

1a. We see job spends on average 15.26 minutes in the system.

Entity Name	Qty Processed	Average Cycle Time (Minutes)	Average VA Time (Minutes)	Average Cost
Call	30151	**15.26**	8.29	0.00

1b. We find that Stage 1 worker is busy 57.56% of time. Stage 2 and Inspection workers are busy around 72% of time.

RESOURCE STATES BY PERCENTAGE

Resource Name	Scheduled Hours	% In Use	% Idle	% Down
Manager	1000	57.56	42.44	0.00
Manager2.1	1000	71.97	28.03	0.00
Manager2.2	1000	71.97	28.03	0.00
Manager2	2000	71.97	28.03	0.00
Manager3.1	1000	71.78	28.22	0.00
Manager3.2	1000	71.78	28.22	0.00
Manager3.3	1000	71.78	28.22	0.00
Manager3	3000	71.78	28.22	0.00

2a. ENTITY SUMMARY (Times in Scoreboard time units)

Entity Name	Qty Processed	Average Cycle Time (Minutes)	Average VA Time (Minutes)	Average Cost
Customer	39773	**2.55**	0.80	0.00

Average customer spends 2.55 minutes in system.

2b. ENTITY SUMMARY (Times in Scoreboard time units)

Entity Name	Qty Processed	Average Cycle Time (Minutes)	Average VA Time (Minutes)	Average Cost
Customer	39757	**1.19**	0.40	0.00

With no wanding it takes an average of 1.19 minutes to go through system.

2c. With extra X-Ray Machine mean time in system is reduced to 1.55 minutes.

ENTITY SUMMARY   (Times in Scoreboard time units)

Entity Name	Qty Processed	Average Cycle Time (Minutes)	Average VA Time (Minutes)	Average Cost
Customer	39825	**1.55**	0.81	0.00

With extra Wand mean time in system is reduced to 1.87 minutes.

ENTITY SUMMARY   (Times in Scoreboard time units)

Entity Name	Qty Processed	Average Cycle Time (Minutes)	Average VA Time (Minutes)	Average Cost
Customer	39972	**1.87**	0.82	0.00

Therefore better to add an X-Ray machine.

Chapter 23 Solutions

See file S23_x_i for problem i of Section 23.x.

Notes on Section 23.4

IMPORTANT! File weibullest.xls presents a better way (using SOLVER) to estimate alpha and beta. Just input mean and sigma and run solver to get correct alpha and beta values. Also in Example 5 problem should say mean = 50 months and sigma = 14 months. In Step 1 formula should be RISKWEIBULL(4, 55.16). This is entered correctly on spreadsheet that came with book. Answer to Example 5 should read

Thus we find in part (a) there is a 99.54% chance that at least one mirror will fail in 60 months or less and only a .46% chance that all mirrors will work for at least 60 months. In part (b) we find that there is a 98.29% chance that all mirrors will fail within 7 years and only a 1.71% chance that at least one mirror will be working for at least 7 years. In part (c) we find that there is a 98% chance that two or more mirrors will be working for 72 months or less and only a 2% chance that two or more mirrors will be working for at least 72 months.

Chapter 24 Solutions

SECTION 24.4

1a. 4,000 cases

1b. After May Prediction for Any Future Month=4,000+.2(4,500-4000) = 4,100.
After June Prediction for Any Future Month=4100+.2(3500-4100) =3980

1c. Simple exponential smoothing gives the most weight to the most recent observation. Thus even though the demand since beginning of May has averaged 4,000 cases, our forecast decreases because the June observation is given more weight than the May observation, and June demand was less than May demand.

2. New Base= $.2 \dfrac{650}{1.3} +.8(440)=452$

New Trend=$.3(452-400)+.7(40)=43.6$

New Summer Seasonality= $.5 \dfrac{650}{452} +.5(1.3)=1.37$

Fall Forecast= $0.8(452+43.6)=396.48$

Winter Forecast=$0.7(452+87.2)=377.44$

11. $L_2 = \dfrac{360}{.95} (.2) + (.8)(300 + 30) = 339.8$

$T_2 = 0.4(339.8 - 300) + 0.6(30) = 33.9$

$S_2 = \dfrac{360}{339.8} (0.5) + 0.5(.95) = 1.00$

4th quarter 1992 demand forecast = $(339.8 + 2(33.9))1.2$ = 489.1 billion dollars 2nd quarter 1993 forecast = $(339.8 + 4(33.9))(1)$ = 475.4 billion dollars.

well as teams.

SECTION 24.5

1. For each professor's payday compute (Actual Customers/Forecasted Customers). Average these ratios. Suppose you obtain 1.3. Then to obtain forecast for a day on which professors are paid, compute a forecast by our basic method, and then multiply this forecast by 1.3.

SECTION 24.8

1a. See file S24_8_1.xls. The beta of GM is the amount by which a 1% increase in the market will increase the return on GM. Thus the beta of the stock will be the coefficient of the independent variable in a regression where the independent variable is S & P return while the dependent variable is GM return. A beta>1, means that a stock is more volatile than the market while 0<beta<1 means that a stock is less volatile than the market. A beta<0 means that a stock goes down when the market goes up. A negative beta is rare (gold stocks have a negative beta; however). Running a regression with GM return as the dependent variable and S & P return as the independent variable yields (see printout) β_0 = 2.89, β_1 = .877, Std Error β_1 = .343 and R^2 = .45. Thus we estimate the beta of GM to be .877.

1b. t = .877/.343 = 2.56. Since $t_{(.025,8)}$ = 2.306 we reject H_o and conclude that the market has a significant effect on GM's return.

1c. R^2 = .45.

1d. 1 - R^2 = .55

1e. $\hat{Y}$= 2.89 + .877(15) = 16.05%

2a. A plot of cumulative production against labor hours for last unit produced looks like (b). Thus we run a regression on the transformed points (ln 1, ln 715), (ln 2, ln517), etc. See file S24_8_2.xls. This yields β_0 = 6.58 and β_1 = -.475. Then $\bar{Y}$ = $e^{6.58}x^{-.475}$. We predict the 800th unit to require $e^{(6.58+.5(.014)^2)}(800)^{-.475}$ =30.1 hours.

2c. We find from file that MAD = 1.43 and s_e = 1.25(1.43) = 1.79 Thus 95% of all predictions should be accurate within 2(1.79) = 3.58 hours.

5a. and 5b. Running a simple linear regression (see file S24_8_5.xls) yields

Monthly Cost = 291.3 + 2.9(units shipped). Plotting the errors for this regression we find that the first ten errors are negative and the last five errors are positive. This indicates that positive autocorrelation might be present.

5c. It seems logical to believe that shipping costs would increase during the trucking strike. To model this we use a dummy variable T, where T = 1 during a month in which there was a trucking strike, and T = 0 otherwise. We now obtain

Monthly Cost = 150.7 + 292T + 3(units shipped). All variables are significant and R^2 improves from .94 to .99. Plotting the errors for this new equation we find that the errors change sign much more often, so the positive autocorrelation has disappeared.

6. Miles driven and Age of the Car are negatively correlated. Thus older cars tend to be driven less. When the computer is told that miles driven is large, this means (for our data set) that the car tends to be a newer car which tends to have low maintenance cost. This explains why a regression with the sole independent variable being miles driven yields a negative coefficient. This example indicates that when an important independent variable is omitted from a regression the estimates of other independent variable coefficients may be biased and inaccurate.

SOLUTIONS TO CHAPTER 24 REVIEW PROBLEMS

1a. Letting Si = 1 if quarter is an i'th quarter of a year and Si = 0 otherwise we obtained the following equation (see file S24rp_1):

$\hat{QUART}$ = -10 -.36PRICE + .00655INCOME - 1.32S1 - 1.33S2 - 2.21S3.
S3 is significant for α = .10, but S1 and S2 are not significant for α = .10. I feel that dummy variables for seasonality are a "package" and you must accept all or none. Thus I included all dummy variables in equation. Errors change sign only twice indicating positive autocorrelation is present. If GNP were included with INCOME we would have multicollinearity. Examining the correlation between the independent variables in the equation appears to indicate that there is no problem with multicollinearity. R^2 = .92 indicates that there is a good fit.

1b. Quarter 4 1980 Prediction = -10 - .36(45) + .00655(10,000) = 39.3 billion pounds. Quarter 1 1981 Prediction = -10 - .36(45) + .00655(10,000) - 1.32 = 37.98 billion pounds.

1c. s_e = 1.56 so 95% of the time we expect our forecasts to be accurate within 3.12 billion pounds.

1d. See file SRP24_1. We used solver to optimize alpha, beta and gamma. s_e = 1.25(1.28) = 1.60, so the regression forecasts are slightly more accurate.